ENGINES OF REBELLION

MARITIME CURRENTS
HISTORY AND ARCHAEOLOGY

SAXON T. BISBEE

ENGINES OF REBELLION

CONFEDERATE IRONCLADS and STEAM ENGINEERING in the AMERICAN CIVIL WAR

THE UNIVERSITY OF ALABAMA PRESS
TUSCALOOSA

The University of Alabama Press
Tuscaloosa, Alabama 35487-0380
uapress.ua.edu

Typeface: Scala Pro

Cover images: *Top two images*, lines of USS *Arctic* as *Lightship No. 8* (see figure 4.1); *middle two images*, plans of *Fredericksburg* (see figure 8.2); and *bottom two images*, plans of Porter's light draft diamond hull centerwheel ironclad (see figure 8.3)
Cover design: David Nees

Library of Congress Cataloging-in-Publication Data

Names: Bisbee, Saxon T., 1986– author.
Title: Engines of rebellion : Confederate ironclads and steam engineering in the American Civil War / Saxon T. Bisbee.
Description: Tuscaloosa, Alabama : The University of Alabama Press, [2018] | Includes bibliographical references and index.
Identifiers: LCCN 2017053877| ISBN 9780817319861 (cloth) | ISBN 9780817391881 (ebook)
Subjects: LCSH: Confederate States of America. Navy—History. | Armored vessels—Confederate States of America—History. | Confederate States of America—History, Naval. | United States—History—Civil War, 1861-1865—Naval operations.
Classification: LCC E596 .B63 2018 | DDC 973.7/13—dc23
LC record available at https://lccn.loc.gov/2017053877

To my *morfar* [the Swedish term for maternal grandfather], Thomas Duane Larson, who embodied all the best aspects of naval officer, engineer, and learned small-town American

CONTENTS

ILLUSTRATIONS

ACKNOWLEDGMENTS

Pondering and researching is one thing; actually writing is quite another. I have many wonderful people to thank for helping me in this regard; I hope I have not forgotten anyone. First and foremost, I thank Brad Rodgers for his patience and for helping greatly to make my work more readable. I also thank Nathan Richards and Wade Dudley, experts in historical maritime archaeology and the history of sea power. I feel especially privileged to have corresponded closely with Bill Still, an expert on the Confederate navy, through all phases of the research and writing process.

I owe a great deal to Bob Holcombe, who next to Dr. Still is probably the world's leading expert on Confederate ironclads. There is a large amount of material in this work that he kindly provided, and the book would have been much less comprehensive without his input. I am also grateful to the staff of the Port Columbus Civil War Naval Museum, especially Bruce Smith, Ken Johnston, Jeff Seymour, and Jerry Franklin. Their hospitality and dedication made my research and stay in Georgia very enjoyable.

Other helpful people were David Rousar, who allowed me to use scans of never-before-seen original machinery plans, and Kazimierz Zygadlo and Bil Ragan, for their generous sharing of obscure images and data. Also invaluable were the staffs of the National Archives; Emory University's Manuscript, Archives, and Rare Book Library; University of North Carolina's Southern Historical Collection; University of Alabama's Special Collections; the Alabama Department of Archives and History; East Carolina University's Joyner Library; the North Carolina Underwater Archaeology Branch; the Virginia Historical Society; the Museum of the Confederacy; the CSS *Neuse* Interpretive Center; and the Mariners' Museum Library.

I thank my family and friends for their love and support on all fronts. A final thanks is also due to Fred Hocker, who initially convinced me to pursue this difficult topic.

A NOTE ON UNITS OF MEASUREMENT

All dimensional units are in feet and inches. Hull lengths and widths, depths, frame dimensions, engine cylinder diameter, and piston stroke were originally conceived, crafted, and noted using that system of measurement and its attendant mindset.

Units of weight are primarily given in tons displacement. This measure represents the amount of water actually displaced by a floating vessel, thus equal to its true weight. Source materials generally make no distinction between the short ton (2,000 pounds) and the long ton (2,240 pounds), although the latter was the primary unit used for calculating displacement.

Tonnage is sometimes given in lieu of displacement; it is a measurement of volume pertaining to all of a ship's enclosed space. Tonnage reflects a vessel's size, not weight. There are a few different types of tonnage measurement (e.g., gross and net), but register tonnage is the only one found in this work. One register ton is equal to the volume of 100 cubic feet.

Knots are nautical units of speed; a knot is simply 1 nautical mile per hour. One knot is equal to about 1.15 miles per hour. Speed in miles per hour is sometimes presented because that measurement could be used by vessels serving on the extensive inland waterways.

Boilers and engines performed work measured in horsepower (HP). Original sources usually make no distinction between the several kinds of specific HP measurement (e.g., indicated, shaft, or brake); thus it is a general term of power generation. One HP equals about 746 watts.

Pounds per square inch (PSI) and revolutions per minute (RPM) are also commonly encountered. The former relates to the vital consideration of ideal operating steam pressure (especially for boilers), and the latter indicates an engine's ideal operating speed (or range of speeds). RPM can also describe the speed at which propellers or paddlewheels are turned.

ENGINES OF REBELLION

ONE

ORIGINS, BACKGROUND, AND TECHNICAL DEVELOPMENTS

Offers are invited by this Department . . . for constructing Engines, Propellors, and boilers for [war] vessels.
—Confederate States Navy Department contract form, 1861

Looking back from a twenty-first-century vantage point, we can readily see that virtually all of modern technological society has its origins in the nineteenth century. Rapid and revolutionary changes occurred in the ways people lived, worked, and waged war. Some of the greatest effects of the technological advances of that century were seen in the world's navies, and from these changes emerged the concept of modern naval engineering and practice. Wooden, sailing line of battleships had seemingly remained little changed for centuries, yet in the mid-nineteenth century they were made obsolete, giving way to vessels armored with iron and steel, equipped with rifled guns, and powered wholly by steam. These powerful weapons were termed *ironclads.*

The acceptance and spread of the armored, self-propelled warship represented a truly revolutionary change in warfare at sea, and many remarked upon it. A particular event in 1862 spectacularly emphasized that process. According to historian James Baxter, "The Elizabethan seadogs who circled the globe with [Sir Francis] Drake might have felt at home in the sailing sloop of war *Cumberland,* as she sank with colors flying on the 8th of March, 1862. Of the five great naval revolutions of the nineteenth century—steam, shell guns, the screw propeller, rifled ordnance, and armor—one only had influenced her design or equipment. Nothing but her heavy battery of 9-inch smooth-bore shell guns would have seemed wholly unfamiliar to the conquerors of the Spanish Armada."[1] The vessel responsible for the *Cumberland*'s demise, CSS *Virginia,* was an ironclad, a product of industrialization in Great Britain, France, Russia, and the United States.

The engagement referred to has become one of the most famous naval battles in world history, and it took place March 8–9, 1862. It was fought in two parts, the first of which (on March 8) resulted in the figurative "death of the wooden warship," whereas the battle on March 9 was the first ever fought between armored steam-driven warships. The two primary participants were the aforementioned CSS *Virginia* (more often remembered by its previous name, the *Merrimack*), and USS *Monitor*. Although the former had once itself been a wooden warship, it was converted to an ironclad and proved the superiority of ironclads over wooden vessels, whereas the *Monitor* in turn rendered the *Virginia* obsolete by its use of a revolving turret. The conflict in which these two vessels participated became known as the Battle of Hampton Roads, fought during the American Civil War.

The Civil War is one of the most scrutinized conflicts in American and world history. This conflict saw four long years (1861–1865) of bloody fighting between Union and Confederate forces. The great land campaigns carried out by Robert E. Lee, William Tecumseh Sherman, and Ulysses S. Grant are very well covered in academic and popular works, but technological development during what is often considered the first modern war has not been examined as extensively. Like all large conflicts, the Civil War became an effective proving ground for new military technologies, including ironclads.[2] Armored vessels marked the peak of technological development during the conflict and indeed became one of the Civil War's most famous aspects.

The development of effective steam propulsion technology was one of the greatest factors in the realization of ironclad warships during the 1850s and 1860s. Unfortunately, serious studies concentrating on the marine propulsion machinery of the Civil War era are quite rare. Confederate ironclads in particular have garnered very little attention regarding their machinery, although the Confederacy was often quite creative in building and obtaining marine power plants. The improved mid- and late-war models built with experience gained from the *Virginia*'s deployment especially deserve consideration.

Not widely appreciated is the fact that technical data for many Confederate ironclads still exists, albeit in scattered locations. Examination of this data offers a very interesting and broad picture of how marine steam machinery was manufactured, obtained, and dispersed to the appropriate vessels throughout the Confederacy.

Although the Confederate States of America ultimately began construction on about fifty ironclads of various types, only twenty-five saw any sort of active service. This work focuses exclusively on those with American-built machinery, offering a detailed look at marine steam engineering practices in both Northern and Southern industry before and during the Civil War.[3] In addition, four nearly completed ironclads are examined in detail because of the high quality of their late-war hull designs and machinery.

In all, twenty-seven vessels—those completed and nearly completed, having received at least a portion of their machinery before destruction—are discussed in this book. There were probably more than this number in reality, but a lack of records prevents any certainty. For instance, whereas the mighty CSS *Mississippi* is well covered, the "Bigbee boats," three or four ironclads under construction on the Tombigbee River of Alabama that were nearly complete at the time of their destruction, are not. Such is the case with most of the unfinished ironclads—there are only small hints and uncertainties. Therefore, although there are many interesting possibilities for future research on both the foreign-built and the unfinished ironclads, this work concentrates only on the finished products, as well as a few of those nearly finished: the *Mississippi*, the *Jackson* (often called the *Muscogee*), the *Milledgeville*, and the *Wilmington*.[4]

The Formation of the Confederate States Navy

The beginning of the Confederate States Navy lay largely in the separate policies of the seceding states. When the Civil War began, several vessels were seized for military service, but the Confederate naval forces of 1861 were a far cry from those that allowed the completion of CSS *Virginia* one year later. Indeed, by and large the Confederate navy ultimately pushed the boundaries of ingenuity and changed the character of war at sea. All this becomes more impressive when one realizes that it had to start from almost nothing, for the navy's first few small and widely scattered vessels were weak and unfit.[5] The fledgling fleet needed the guiding hand of an able administrator, and it soon received one.

The extraordinary resistance shown against the Union navy, the skilled development of the ironclad, commerce raiding, and mine warfare programs were all mostly a result of one man: Stephen Russell Mallory. This capable and innovative administrator served for the entire Civil War period as the Confederate secretary of the navy, and he built it into a modern and effective military service by 1865. By that time, the ironclad program was the most important component of that service. To fully appreciate Mallory's achievements in this area, we must look in detail at his first major challenge: actually organizing a navy.[6]

Little is known about the future secretary's early life, and even the exact year of his birth is uncertain. Several birth dates, ranging from 1811 to 1814, are stated in various works, but the consensus now is that he was born around 1813.[7] It is established that by the age of nine Mallory was living on Key West, where he spent his entire youth and early adulthood. He became quite familiar with the sea and ships even while practicing as a

Florida lawyer, and his intelligence and reputation helped him acquire a US Senate seat in 1851. In 1853 he began his first major appointment, as chairman of the Senate Naval Affairs Committee, a post foreshadowing his office during the Civil War.[8] The Florida senator, always interested in ships and naval affairs, executed his duties zealously. This had the unfortunate effect of gaining Mallory few friends in naval service because of his support of unpopular legislation, such as reinstating the use of corporal punishment and the formation of the Naval Retiring Board for older officers. Nevertheless, many of the bills he supported passed, and Mallory remained a capable chairman until the advent of the Secession Crisis.

During the crisis, Senator Mallory, like many others, was torn between his conflicting duties to his home state and the federal government. In March 1861 this issue became null when the senators from the seceded states were removed from the Senate roster. Mallory himself, along with several other Southern politicians, had actually left office in February.[9] Soon the future Confederate naval secretary was heading for Montgomery, Alabama, where both the Confederate government and Mallory's most challenging project were taking shape.

Montgomery was chosen as the first Confederate capital simply out of necessity. It was far from Union lines, and the Provisional Congress of the Confederate States and its standing committees, among them the Naval Affairs Committee, first met there on February 19, 1861. The previous day, Senator Jefferson Davis of Mississippi was sworn in as president. (Throughout his career Davis never showed much interest in naval affairs, and throughout the war he seemed to relegate them to secondary importance.) Matters continued to move quickly, and the Confederate Navy Department was established on February 20. Five days later, President Davis nominated Stephen Mallory to his cabinet as naval secretary. The organization of a true navy soon began.[10]

Following a revolutionary pattern similar to that of eighty years before, the seceding states quickly set about establishing small state navies for local protection. The new organization under Mallory ultimately inherited little from these ragtag forces, for the only vessels the seceded states were able to use were small steamers or other secondhand vessels. South Carolina's navy was certainly the best, although it was still pitifully small, composed of only the old sailing revenue cutter *Aiken* (one gun); the tugboat *James Gray* (rechristened the *Lady Davis* in honor of the new Confederate president's wife); the little steamers *Catawba*, *Gordon*, and *Seabrook* (each hastily fitted with a small-caliber gun); and a few small sailing boats formerly in the Lighthouse Service.[11]

Mississippi had no navy at all, choosing instead to concentrate on land defense, and Mallory's home state of Florida had only a small ex–US

Coast Survey schooner. Alabama seized a revenue cutter and a US Lighthouse Board tugboat; Georgia acquired two sidewheelers, the *Savannah* (ex-*Everglade*) and the *Huntress;* and Louisiana and Texas had only three small revenue cutters between them.[12] This was the entirety of the Confederate vessels obtained in the first round of secession, but Mallory later gained the large steamers *Patrick Henry* and *Jamestown*, the tugboat *Teaser*, and four smaller vessels when Virginia and North Carolina seceded. Even when fully armed and outfitted, however, the armament of all the state navy vessels combined amounted to no more than twenty guns, not even equivalent to the armament of a single US sloop of war.[13]

Despite the enormous organizational tasks they faced, Secretary Mallory and other Confederate leaders understood that a war with the states remaining in the Union was inevitable and that they must act quickly. Therefore, with what few vessels he had, Mallory set about organizing a naval infrastructure. When the Confederate capital was relocated to Richmond, Virginia, the secretary and his assistants set up the Navy Department offices at the old Mechanic's Institute on Ninth Street. Four bureaus were established: the Office of Orders and Detail, the Office of Ordnance and Hydrography, the Office of Provisions and Clothing, and the Office of Medicine and Surgery. In addition, the Confederate Marine Corps and the positions of chief naval constructor and engineer in chief were subsequently created. All functioned very similarly to the venerable departments of the US Navy.[14]

Starting with only that handful of vessels gained from the absorbed state navies, Mallory quickly implemented a mass building strategy to counter the numerically superior Union navy. Because of his several years as chairman of the Senate Naval Affairs Committee and his great interest in new shipbuilding technologies, Secretary Mallory was keenly aware of the latest developments in naval armaments and armored vessels. He also supported the development of new methods of warfare, such as the use of mines (then called torpedoes) and commerce raiding. Although these weapons had occasionally been deployed in previous conflicts, Mallory's adoption of them using the modern technology of the 1860s allowed their use on a scale never before seen in naval war.[15] The nascent Confederate navy stood poised to overwhelm the aging wooden ships of the Union with ultramodern cruisers and ironclads—quality against quantity, if only enough time could be had to fully realize the construction programs.

The Confederate navy secretary's approach to new construction consisted of the following: a large number of small, simple-to-construct wooden steam gunboats would be built, allowing time for a smaller number of larger and more powerful ironclads to be built in conjunction with the construction or obtaining of cruisers, often from foreign contractors, for commerce raiding. Whereas the commerce raiders were primarily built abroad because of the Confederacy's strong ties with Great Britain, the wooden gunboats and ironclads were mostly contracted out to private builders throughout the Confederate States. By July 1861 the gunboat and ironclad programs were in full swing, and the first major commerce raiders, led by CSS *Sumter*, had begun their first depredations against Union merchant vessels.[16] The Confederate navy, though hastily organized with no important base to build on, had successfully begun operations.

The Confederate Ironclad Program

The origin and first steps of ironclad construction were envisioned by Stephen Mallory and executed by a few select men. It is well-known that the secretary was long fascinated by the new rifled ordnance developed in Europe and the utilization of steam-driven ironclads. He therefore set out to enact a program centered on these two new weapons. Whereas armored floating batteries were first tested in action during the Crimean War, the new technologies of explosive shells and rifled ordnance had not been combined with armor on steam vessels by the 1860s. Even though nearly one hundred ironclads were built or being built in Europe by the Civil War, they did not utilize rifled guns and still clung to auxiliary sail power. Although ships like HMS *Warrior* were certainly powerful, their place in the battle line was not yet firmly established. In contrast, Mallory intended his ironclads to be an ultimate frontline weapon against the Union.

As the former chairman of the Senate Naval Affairs Committee, Mallory had experience in funding ironclad construction projects, and he was soon successful in convincing the Confederate Congress to raise two million dollars for purchasing one or two European ironclads.[17] His view of ironclads as the future of naval technology proved prophetic: "I regard the possession of an iron-armored ship as a matter of the first necessity. . . . If to cope with them [Union navy ships] upon the sea, we follow their example and build wooden ships, we shall have to construct several at one time. . . . But inequality of numbers may be compensated by invulnerability, and thus not only does economy but naval success dictate the wisdom and expediency of iron against wood."[18]

This began the Confederate ironclad program, both at home and abroad, for Mallory also began a massive building program for ironclads within the Southern states. He was aided by several men of like mind in those early days of 1861. The most important among them were riverboat builder E. C. Murray of Kentucky, later the builder of CSS *Louisiana*; Lieutenant John M. Brooke, an ordnance expert and later the head of the Office of Ordnance and Hydrography; and Naval Constructor John L. Porter, later the chief naval constructor and the designer of nearly all Confederate ironclad types.[19] Although Porter had designed an ironclad as far back as 1846 and Murray had proposed a design in April 1861, it was ultimately Brooke's ideas that were first incorporated into the type of ironclad that Mallory desired. Many of Porter's concepts of 1846 were also used, as well as several new ones he developed during the design process for a new Confederate ironclad. By June 1861 the process was well under way of constructing what Mallory saw as an ideal design.

Mallory, Brooke, Porter, and Engineer in Chief William P. Williamson jointly determined that constructing the Brooke-Porter ironclad would be much easier if an existing vessel was converted. For this purpose the submerged, partially burned hulk of the former US steam frigate *Merrimack* was selected. It lay at the Confederacy's largest and most valuable naval facility: the Gosport Navy Yard in Norfolk, Virginia. The *Merrimack*'s power plant was largely intact, allowing valuable time and money to be saved by not having to construct new machinery. Therefore, a burned-out steam frigate became the prototype Confederate ironclad.

The *Merrimack* (renamed CSS *Virginia*) was not, however, the first armored vessel completed in American waters. A small, privately built ironclad, the *Manassas* (the ex–*Enoch Train*) was finished a year before in Louisiana with an unconventional design. It was protected by a rounded shell of iron plate, containing room for only one forward firing gun, and was the only Confederate ironclad not to use an angular casemate. The *Manassas* ultimately remained more of a Mississippian curiosity during its brief career than a truly effective fighting vessel.[20] Therefore it was the *Virginia*'s design—utilizing the semisubmerged ends, ram bow, and inclined laminated armor casemate developed by Brooke and Porter—and not the *Manassas*'s iron shell configuration that set the standard for later construction.

The *Virginia* was finished as much as possible in March 1862 and rushed into its fateful encounter with Union blockading forces in the Battle of Hampton Roads on March 8, from which it emerged victorious. Even more famous was the fight with the new Union ironclad USS *Monitor* the following day, on which much has been written. The *Virginia* became the first Confederate ironclad to see action against another armored ship, and it set the trend toward the standard casemated harbor defense ironclads designed by Porter throughout the war. The casemate design was effective and simple to construct, and it remained the most obvious feature of the Confederacy's homebuilt ironclads.[21] Although later armored vessels were significantly scaled down from the *Virginia*'s massive size, they all retained the same general layout.

The homebuilt ironclad program as first envisioned by Mallory ultimately became the Confederacy's largest shipbuilding program, despite important shifts in focus. Early ironclads such as the *Virginia*, the *Arkansas*, the *Mississippi*, the *Louisiana*, and the *Georgia* were built according to large, heavily armed, unconventional designs. This was the result of an unclear offensive strategy that required these ships to operate in shallow rivers and harbors as well as at sea against federal blockaders.[22] The inability of the Confederacy to efficiently build and utilize vessels with such capabilities was made apparent to Mallory's administration only after the catastrophic losses of all these vessels except the *Georgia*, which from the outset could be used only as a floating battery. Thereafter, beginning in 1862, Confederate naval strategy shifted to the construction of smaller harbor defense ironclads of somewhat standardized design.

In the end, Secretary Mallory renewed his attempts to build or procure ironclads in Europe, but only one, CSS *Stonewall*, was actually delivered into Confederate hands. That vessel was commissioned too late to take active part in the war, and it did not reach western Atlantic waters until after General Lee's surrender in April 1865. Although the European ironclad efforts can be considered a failure, Mallory's domestic program ultimately unfolded much as he had proposed in May 1861. By the end of the Civil War the ironclad program was the largest naval construction effort in the Confederacy—the early wooden gunboat program had been entirely superseded—and the keels of more than fifty ironclads were ultimately laid.[23] Of these, twenty-five were commissioned into active service. The ironclads helped greatly in delaying the Union takeover of Confederate port cities, despite their low number and success rate in timely construction and effective machinery. Mallory's strong grasp of and appreciation for new technology made the miniscule Confederate navy of 1861 into a pioneering and technologically advanced force to be reckoned with by 1865.

Confederate Ironclad Hull Types

Confederate ironclads can ultimately be grouped into five categories based on hull type: conversions, early nonstandard types, standard hull designs, diamond hull designs, and late-war types.[24] All twenty-seven Confederate ironclads discussed in this work fall under these categories. The conversions were the *Manassas*, the *Virginia*, and the *Baltic*. The early nonstandard types were the *Louisiana*, the *Arkansas*, the *Georgia*, and the *Mississippi*. The standard hull designs were the *Richmond* class (the *Richmond*, the *Chicora*, the *Palmetto State*, the *North Carolina*, the *Raleigh*, and the *Savannah*), the *Tennessee* class (the *Tennessee* and the *Columbia*), the vessels designed by Acting Naval Constructor William Graves (the *Charleston* and the *Virginia II*), and the large sidewheelers (the *Nashville*).

The diamond hull designs were the *Tuscaloosa*, the *Huntsville*, the *Albemarle*, the *Neuse*, the *Fredericksburg*, the *Missouri*, and the *Jackson*. Finally, the late-war ironclads were the *Milledgeville* and the *Wilmington*. All these groups contained other ironclads that were never finished or even named, in some cases, and other groups contained vessels that either were foreign-built or contained foreign-built machinery, which is not discussed in this work.

Some vessels were never completed but came close enough that significant details are known; these were the *Mississippi*, the *Jackson*, the *Milledgeville*, and the *Wilmington*. In addition, some doubt remains about the exact hull configurations of the *Nashville*, which is for now listed as a standard hull, and the *Tuscaloosa* and the *Huntsville*, which are for now listed as diamond hulls.[25] Fortunately, the latter two ironclads are preserved in the archaeological record, making the future positive identification of their hull structures likely.

An explanation of the exact configurations of the different types permits a better understanding of their machinery requirements and layout. The converted ironclads varied greatly: the *Manassas* was formerly the twin-screw towboat *Enoch Train;* the *Virginia* was formerly USS *Merrimack*, a single-screw auxiliary steam frigate; and the *Baltic* was formerly a sidewheel towboat and cotton lighter.[26] Among the early nonstandard ironclads, the *Arkansas*'s design combined elements of both riverboats and coastal steamers in a twin-screw layout, whereas the *Mississippi*'s layout reflected the intention of expediting construction time and employing unskilled labor. It was, accordingly, utterly devoid of shiplike curves. The *Mississippi* was also one of the first triple-screw vessels constructed.[27]

The *Louisiana* was similar to the *Mississippi* in the fact that its hull contained few or no curves, but its power plant was very unusual, consisting of two center paddlewheels mounted one behind the other along with two small propellers for steering. Almost nothing is currently known about the *Georgia*'s hull design other than that it must have been simple and was provided with two propellers.[28]

The standard hull ironclads are, appropriately, much more easily grouped. The six vessels of the *Richmond* class, though differing slightly in casemate and pilothouse layout, employed the same hull design (fig. 1.1). All were driven by a single 10-foot diameter screw.[29] Similarly, the *Tennessee* class vessels employed a standard hull form but with greatly different casemate and armor layouts. They were single screw, although the *Columbia* was originally planned as a twin-screw vessel.

The ironclads designed by William Graves, the *Charleston* and the *Virginia II*, mostly followed the same layout as the previous classes, but they were larger with more rounded bilges. These vessels were also fit-

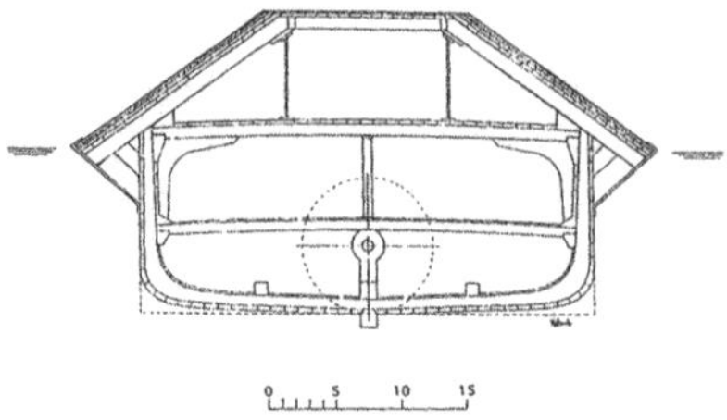

Figure 1.1. Midship section of a *Richmond* class ironclad. The flat bottom, sharp turn of the bilge, and vertical hull sides with the built-on armored knuckle at the waterline are characteristic. The knuckle was an important and distinguishing feature of Confederate ironclad construction, for it allowed the deflective angle of the casemate to continue to the water's edge while also providing a solid defense for the vulnerable upper hull against a low-angled shot or ramming. Note also the large single screw.

ted with a single propeller and carried rams of an unusual protruding design, possibly for penetrating the unarmored hulls of monitors under the armor belt.[30]

As previously stated, the exact hull configuration of the *Nashville* is unknown, but it appears to have been of the standard type.[31] This ironclad was a massive vessel propelled by sidewheels and was the only Confederate ironclad to have equipped sidewheels besides the converted *Baltic.*

Following the more traditional hull layout of the standard types, the diamond hull ironclads utilized a design intended for ease of construction and shallow draft.[32] As many curves as possible were eliminated from the hull, resulting in a hexagonal, or diamond-shaped, midship section (fig. 1.2). The diamond hull ironclads differed based on size and propulsion. Almost nothing is known about the exact layout of the *Tuscaloosa* and the *Huntsville,* but they are thought to have been twin screw.[33] The *Albemarle,* the *Neuse,* and the *Fredericksburg* employed twin-screw propulsion, whereas the *Missouri* and the *Jackson* were designed for a center paddlewheel. The *Jackson* was altered in the course of construction to mount twin screws, making the *Missouri* the only completed Confederate centerwheel ironclad.[34]

The final two ironclads for which significant engineering data exist were never finished and differed greatly in design. The *Milledgeville* was an improved version of the standard hull type, employing twin screws and a greatly improved armor configuration. Perhaps the most important difference was that it had a lighter draft than the standard hulls.[35]

The *Wilmington* was one of the very last ironclads designed in the Confederacy. From the outset it appears to have been intended as a blockade breaker. In its cross-section this ironclad reflects similar improvements in the standard hull form as employed in the *Milledgeville,* but it sports a deeper draft and very fine lines.[36] Also like the *Milledgeville,* the *Wilmington* was driven by twin screws and appears to have lacked a ram.

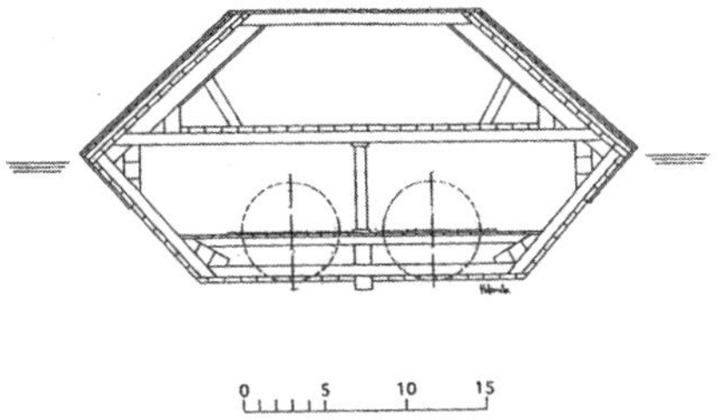

Figure 1.2. Midship section of the *Albemarle*. Note the twin screws, shallower draft, smaller size, and lack of curvature compared to other hull forms.

Neither of these two ironclads was completed, but their designs certainly represented great improvements in hull and machinery layout.

The evolution of Confederate ironclad hull forms reflects a continual trend toward better, cheaper, and more easily constructed vessels. Secretary Mallory and the Navy Department made a great effort to improve the basic ironclad model established by the *Virginia*.[37] The acquisition or construction of steam machinery played an important part in the improvement process. Though less successful in the long run in providing totally reliable engines, the Confederate ironclad program still achieved notable successes in the field of marine steam engineering.

The Marine Steam Engines of the Ironclads

There were many different types of marine engines and boilers in use in the industrialized nations by the 1860s, but only a few were installed on Confederate ironclads. The three main types of engine used were all of the direct-acting type: horizontal, vertical inverted, and inclined (or riverboat type). Direct-acting steam engines, like virtually all other engines during the first half of the 1800s, were usually single cylinder, simple expansion, and double acting. *Simple expansion* denotes the use of steam pressure at a given volume and force in one cylinder only or in multiple cylinders separately. *Double acting* denotes the use of steam to move the piston in both directions in the cylinder alternately. No matter what their arrangement, direct-acting engines worked the same way: the cylinder was positioned to transmit power directly to the crankshaft via the piston rod.[38]

Horizontal direct-acting engines were the most popular type for screw warships during the Civil War period. This arrangement offered low clearance for beneath the waterline protection while also retaining an easily accessible layout. Its inherent disadvantage lay in the extremely short stroke (which could cause stoppages on the top and bottom dead centers). A lack of athwartships space in the bottom of a vessel's hull could also be a problem, exacerbated especially in deep-bottom seagoing vessels. Shallow-draft flat-bottom warships such as the Confederate

Figure 1.3. The two horizontal direct-acting engines of CSS *Chattahoochee*. Each engine turned one screw on the opposite side of the hull from the cylinder head. The broken remains of the valve chests can be seen on top of each cylinder, and the round condenser for the forward engine is visible just in front of its cylinder. The connecting rod from the forward engine to the propeller shaft is missing; all the machinery was smashed with sledgehammers before the *Chattahoochee* was scuttled.

ironclads were much better suited. Indeed, the horizontal direct-acting engine was the most common purpose-built engine used by the Confederate navy. The only known extant example of this engine type in the United States is the set of twin engines in the wooden gunboat CSS *Chattahoochee* (fig. 1.3). About 30 feet of the scuttled gunboat's stern, including the machinery, was raised from Georgia's Chattahoochee River in 1964. The engines are now at the Port Columbus Civil War Naval Museum, a rare surviving example of Civil War steam technology.[39]

Another very common type of propeller engine in the 1860s was the single-cylinder vertical direct-acting variety. The configuration was simple, reliable, and often quite compact, leading to its widespread utilization by small merchant vessels and tugboats. Although the engine was originally designed for turning paddlewheels, it was found that inverting the cylinder over the crankshaft instead of under it provided an even better arrangement, especially for turning a propeller shaft. In this vertical inverted form, the cylinder was placed over the shaft with sufficient room for the piston and connecting rods to turn the cranks. The resulting system was quite accessible and allowed for direct observation and simple maintenance.[40] Its basic components did not change for many decades. A vast majority of the screw tugboats of the Civil War period utilized this engine, and it found its way onto a number of Confederate ironclads. An

Figure 1.4. Early 1900s single-cylinder vertical inverted direct-acting engine for a screw-propelled harbor tugboat. With a cylinder diameter of 18 inches and a stroke of 22 inches, it rated 300 horsepower. The cylinder can be seen at the top, with the valve chest just to the right of it, and the main steam line going into it. The piston and connecting rods are linked below the cylinder but are mostly obscured here by the framing. A small section of the propeller shaft with a thrust bearing can be seen in front of the workmen.

example can be seen on an early twentieth-century New Orleans tugboat (fig. 1.4). This engine hardly differed from the engines utilized by several Confederate ironclads fifty years earlier.

An important and interesting outgrowth of the direct-acting layout, in which the engine cylinders were inclined fore and aft, was commonly used for turning paddlewheels in mid-nineteenth-century American shipping. The design for riparian vessels particularly came to be epitomized by the western river steamboat (fig. 1.5). The shallow waters on which those boats ran favored light, flexible construction and small but powerful engines.

Riverboat engines utilized the same components as other direct-acting forms but required some specialized adaptations. A river steamer's long and flexible but very shallow hull required a large area of even weight distribution and a long, smooth stroke, but other changes, including a significant increase in operating pressure, were made purely for economic reasons.

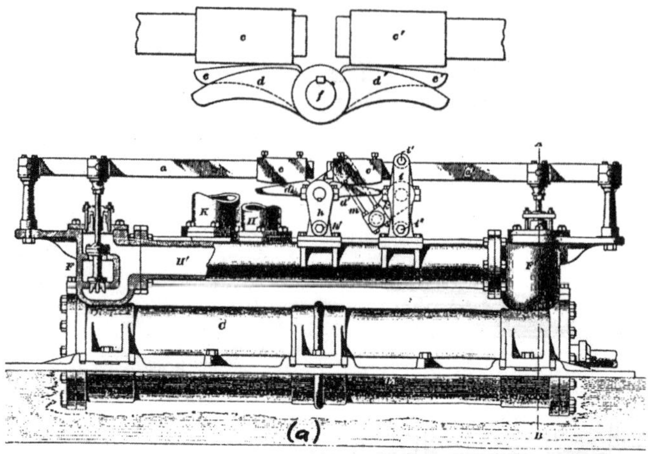

Figure 1.5. A typical western river steamboat engine. In this side view, cylinder *C* is mounted horizontally directly to the vessel's structure through large wooden mounting blocks at the bottom, with the valve apparatus *F, F', H,* and *H'* on top. The cylinder head is at the left, the piston rod at the right, and the valves *F* and *F'* on top at either end. The valve levers *a* and *a'* and wipers *d* and *d'* impart alternating motion to the valves for inletting and exhausting steam to the cylinder via linkages *m, t,* and *t'* to eccentrics on the main crankshaft (not shown). A close-up of the level-wiper interaction at the top shows how the oscillating motion imparted to the centrally pivoted wipers *d, d', e,* and *e'* by the eccentrics alternately raises and lowers the levers *c* and *c'*, which alternately raise and lower the inlet and exhaust valves of the cylinder, allowing steam to move the piston inside. As virtually all steam engines by this time were double acting, there were two valves at each end of the cylinder, one for inlet steam and one for exhaust.

Most important, the heavily forested banks of the Mississippi River system offered nearly unlimited fuel supplies, and this, combined with the desire for a mechanically simple but rugged engine, resulted in the typical riverboat type. This design incorporated a cylinder on each side of the hull driving a crank on the sternwheel (or sidewheel) with a very long wooden connecting rod called a Pittman.[41]

Though influenced by other factors, riverboat engines evolved largely through practical experience, with little thought given to the standardization of power plants or to formally trained engineers until after the Civil War.[42]

The simple and powerful western river engine was fitted to virtually every one of the thousands of paddlewheel steamboats plying the Mississippi River Basin and other rivers in the American South and West. It was utilized by the hundreds of steamers pressed into Union or Confederate war service, and several sets of machinery were salvaged for use on new vessels, including ironclads. As we will see in a later chapter, western riverboat engines and boilers were even adapted by the Confederates to turn propellers.

An important outlier not driven by direct-acting machinery was USS

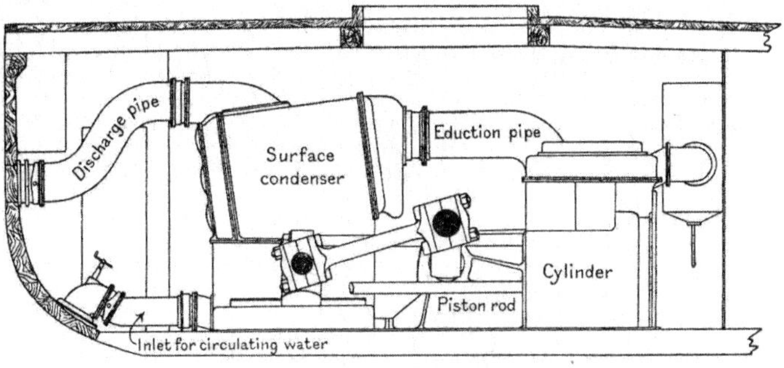

Figure 1.6. A return connecting rod engine, shown by way of a vessel midship section. The main crankshaft is just to the left of the cylinder. In order to clear the shaft, two piston rods were used to connect to the crosshead on the other side, just under the surface condenser in this view. The connecting rod then returned to the crankshaft to impart its motion. This setup was suitable to large single-screw vessels like USS *Merrimack*.

Merrimack, or CSS *Virginia*, which utilized return connecting rod (or back-acting) engines (fig. 1.6). This type of engine was originally used for turning paddlewheels, but its arrangement lent itself quite readily to turning screws. It was also an attempt at a horizontal layout without the problems created by short connecting rods. Henry Maudslay of Great Britain appears to have been the originator of this design in the 1840s. Although the return connecting rod engine benefited from a long stroke, it still took up a considerable amount of space, was relatively complex, and required additional components.[43] In its actual layout, one piston rod split into two or more arms was utilized. These passed down each side of the cylinder and were joined by crossheads. A wishbone-shaped connecting rod extended upward from the short crossheads to turn the overhead crankshaft.[44] The engine was often called *back acting* for this reason. The *Virginia*'s engines were typical of this type except for their size.

Confederate ironclads also had to make do with adapted machinery technologies. At the time of the Civil War, a few of the older engine forms originally designed for paddlewheel propulsion were still being utilized to turn screws. Methods of adaptation were necessary because of the relatively slow turning rate of paddlewheels: an engine had to develop only about 25 revolutions per minute (RPM) to make a paddlewheel practical. Propellers, in contrast, require at least 50 RPM to function properly. In the first days of screw propulsion, therefore, the most common way to achieve the desired speed was through gearing.[45]

The use of geared engines, particularly in warships, was largely a stopgap measure until better systems were developed. In the case of a few Confederate ironclads, the only available machinery was not ideal for screw propulsion, so gearing was used by necessity even at that late time. Geared engines had many disadvantages, such as excessive weight, noise generation, and the occupation of more space than regular power plants.[46]

Before the widespread use of direct-acting engines for screw propulsion, it was considered too difficult, even in short-stroke engines, to obtain 50 or more RPM; consequently, it was the universal custom to use multiplying cogwheels to achieve the desired revolutions.[47] The wheels were often fitted with wooden teeth when attached to the engine shaft, and with iron pinions when attached to the propeller shaft. In most cases the engine shaft was placed across the hull (the engines were placed fore and aft, and their cranks were directly coupled to either end of the shaft as in a paddlewheel arrangement), and the cogwheel was positioned so that it meshed at a 90-degree angle with the cogwheel on the propeller shaft. During the Civil War several Confederate ironclads, including the *Tennessee*, the *Tuscaloosa*, the *Huntsville*, the *Albemarle*, the *Neuse*, and the *Wilmington*, were fitted with geared machinery of various layouts. These systems were cumbersome but ultimately functioned quite well in nearly all cases.

The Marine Boilers of the Ironclads

Boilers are the heart of a steam power plant and, like engines, initially took many different forms. For marine applications, however, two dominant types emerged by the 1860s: the tubular boiler and the return flue boiler. All Confederate ironclads were powered by these, more often than not taken from previous vessels, but all in all the boilers proved less troublesome than the engines. Boiler technology development had stabilized somewhat by the time of the Civil War, and engineers, both naval and merchant, were very familiar with the main types. Although there was no consensus on what boiler was best, 1860s-era boilers were nearing an ideal level of efficiency, reliability, and power generation.

As with steam engines, there were several different boiler types in use by the 1860s, but two basic designs, return flue and tubular, were most prominent for maritime use. Both burned coal or wood. A flue boiler utilized one or more large passages (flues) through the length of the boiler to return the exhaust gases passing from the furnace to the smokestack. The passage of the exhaust through the boiler heated the surrounding water encased in the boiler shell. A tubular boiler worked on the same concept but replaced a few large flues with dozens or hundreds of smaller tubes. Because more small tubes offered a similar heating surface area in a smaller space than a few larger flues, the tubular boiler was more compact, with similar vertical and horizontal dimensions, whereas the flue boiler was usually a

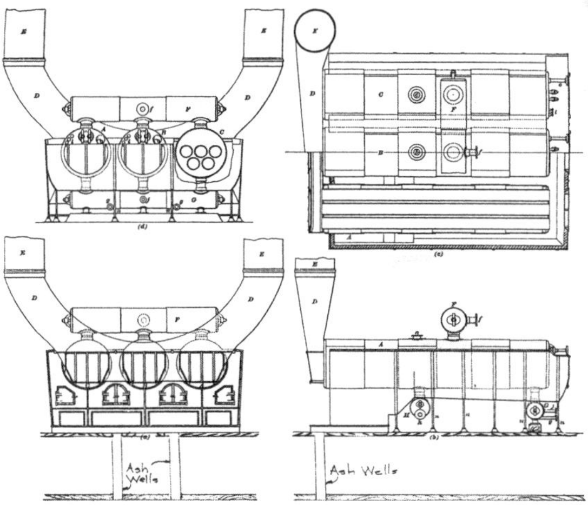

Figure 1.7. Typical return flue boiler layout of western river steamboats. This particular example features a battery of three boilers labeled *A*, *B*, and *C* with five flues each. At the lower left is a front view of the setup: above the ash wells (which pass down through the boat's hull for the disposal of waste ash), the furnace doors can be seen. Immediately above those, on the boiler fronts, are the smokebox doors for access to the uptakes; the uptakes, or breechings, are labeled *D*, and the bases of the smokestacks are labeled *E*. Steam drum *F* collects steam for delivery to the engines. At the upper left is an aft view emphasizing the steam drum at the top and mud drums *G* and *H* at the bottom, with a cutaway of the right boiler shell, showing the flues, which are surrounded with water encased in the boiler shell. The mud drum collects sediment and minerals precipitated out of the boiler water. The top view at the upper right is another cutaway of the lower boiler shell to show the flues, and it demonstrates how the boiler bank is encased in fireproof materials *c* to allow for the initial passage of the exhaust gases underneath and behind the boiler shells before entering the flues. The lowercase letters *a*, *b*, *f*, *g*, *h*, *j*, and *k* represent the manhole for internal maintenance access, the bedplate, the main steam line to the engines, three steam lines to auxiliary equipment, and the boiler support rods, respectively. It is important to remember that all boilers, whatever their type, utilized most of the components listed here.

very long cylinder. At the time of the Civil War, the flue boiler was still more common; the tubular boiler was only just coming into its own.

The flue boiler had its own specific advantages and disadvantages. It was compact vertically, had a large heating surface, and burned fuel more efficiently. It was very common for shipping interests, especially riverboats. Unfortunately, a flue boiler weighed more, was usually very long, and required more care because of the liability of the flues becoming overheated and collapsing if the regular supply of water was neglected.[48] A collapsed flue was the result of the water level in the boiler being too low; although the water itself was constantly heated to the boiling point, its very presence as a liquid under pressure kept the metal of the flues from melting. If a hole was melted through, an almost instantaneous explosion occurred as superheated water was released from pressure, bursting the boiler, scattering the fires, and often destroying the vessel as it flashed to steam. A buildup of pressure-cooked mud and minerals (called *scale*) on

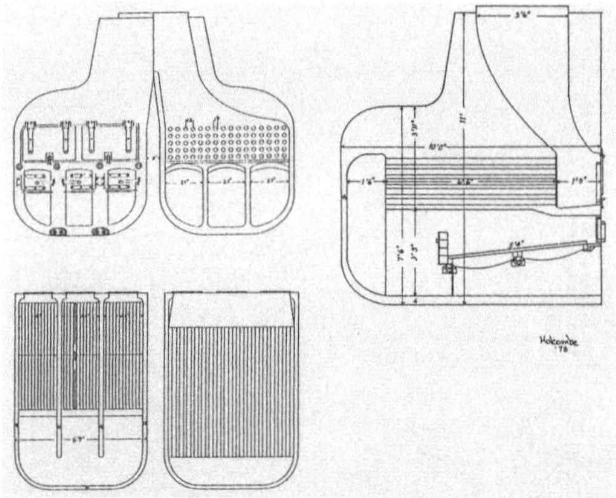

Figure 1.8. A pair of low-pressure marine horizontal firetube boilers, used on several Confederate ironclads. The front view at the upper left demonstrates both an intact aspect on the left and a cutaway on the right. This particular example sports three furnaces, above which are two large smokebox doors for uptake access. The cutaway demonstrates the firetube ends, whereas the side view at the upper right shows the arrangement more clearly. The furnace with its fire grate is at the bottom, and the passage for exhaust gases around and up the back of the boiler into the firetubes (which act as many small flues) can be seen. On exiting the firetubes, the exhaust gases are funneled up the smokestack. The boiler water is encased within the area surrounding and above the firetubes. Many boilers of this type were manufactured at the Columbus Naval Iron Works in Georgia and were used in all the *Richmond* class ironclads.

the outside of the flues, even when solidly encased in water, could cause the same reaction. Constant maintenance was therefore necessary, especially on vessels that utilized the fresh but muddy water of harbors and rivers, such as western river steamboats.

Western river steamboat boilers (fig. 1.7) in particular were a marvel of practicality and power generation. They were of the return flue type, almost always being employed in a battery of two or more or, in at least one case, sixteen.[49] The average size of these boilers was about 36 inches in diameter and 28 feet long. The length helped to spread the weight evenly over a riverboat's flexible hull, and the low profile was useful for saving vertical space between decks. A very high steam pressure of 100 pounds per square inch (PSI) or more was utilized, in marked contrast to the common practice on the coasts or in oceangoing ships. The majority of the western riverboats burned wood, although coal was used when it could be obtained. Because of this versatility, the high power-to-weight ratio, and the general abundance of used sets, western river–style boilers became probably the most common form of boiler employed by Confederate ironclads.

The tubular boiler, meanwhile, possessed many advantages of its own but was more complex, making it more expensive to build and maintain. This form of boiler occupied less space generally, required less fuel, and was not liable to collapse, because of the small size of the individual tubes

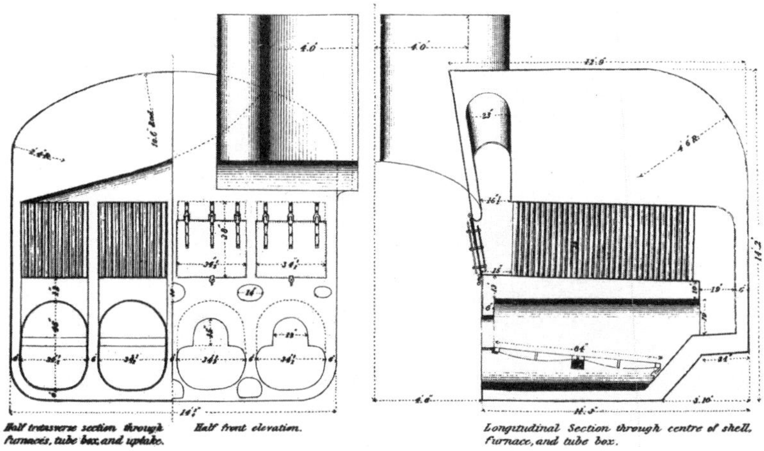

Figure 1.9. Martin vertical watertube boiler of CSS *Virginia*, ex–USS *Merrimack*. The front view at the left demonstrates four furnaces with the vertical watertubes above. The side view cutaway at the right clearly shows a furnace with its fire grate at the bottom and the passage for exhaust gases as they travel up and among the watertubes on their way to the smokestack uptakes. The watertubes connect the upper and lower portions of the boiler water storage within the boiler shell.

(although replacing failed tubes could be a difficult procedure, requiring not just hauling the fires but also emptying the boiler of all its water). By the end of the nineteenth century, the tubular type of boiler superseded all other forms in oceangoing vessels, but this important transition did not begin to occur until after the Civil War.

Tubular boilers ultimately took two forms: the firetube, in which the many small tubes, surrounded by water, acted as miniature flues for the furnace exhaust gases (fig. 1.8); and the watertube, in which exhaust gases surrounded the tubes, which contained the water (fig. 1.9). Both types had horizontal and vertical forms. The compact horizontal form was most common. The vertical design was often employed in small auxiliary systems but was also found on some major warships, including USS *Merrimack*. The average size was about 10 feet in diameter by 15 feet long. All forms of the tubular boiler were more efficient at transmitting heat to the contained water but were not as good for perfect combustion as the flue boiler.[50] Like many oceangoing systems, they were commonly low pressure.

The terms *high pressure* and *low pressure* in relation to engines have already been mentioned in passing but must be fully explained in discussing boilers, because the boiler is the pressure vessel that actually produces the power by which a steam engine acts. Low-pressure systems generally operate at or slightly above atmospheric pressure (approximately 14 PSI). For most of the nineteenth century, oceangoing vessels, both merchant and naval, operated on low pressure.

The terms relating to pressure, however, were liable to change as better-quality construction, the understanding of physical principles, and improved designs were incorporated into marine boilers. In marine engineering the term *high pressure* was applied to steam at even 15 PSI. It may be concluded, then, that 5 PSI was considered ordinary steam pressure in marine boilers in the 1830s, 10 PSI in the 1840s, and 20 PSI in the 1850s. Vessels using steam at any pressure in excess of these were considered to be using high-pressure steam.[51]

These so-called high pressures paled in comparison to the 100 PSI or more employed by western river steamboats, as previously mentioned, but professional engineers, especially naval ones, tended to be cautious. Their careful, often scientific approach to rapidly changing technology stood in stark contrast to the no-nonsense, experience-based attitude of river steamboat engineers. Nevertheless, all had to cope with the demands of their boilers, engines, and lesser steam systems in much the same ways.

Auxiliary Equipment

An obscure but vital part of any steamship's power plant was its auxiliary systems, which drew steam from the main boilers to perform a number of lesser tasks. Especially on ships, these duties revolved mostly around pumping. There were a number of auxiliary systems in use on steam vessels during the Civil War era, but they remained highly nonstandardized for many years. This usually reflected the business practices of the western rivers, where the adoption of new technology was often seen as unnecessary. Nevertheless, important advancements were made by the 1860s.

There were some important differences in auxiliary equipment between riverboats and oceangoing vessels. The latter, including warships, generally utilized low-pressure engines, as already discussed. A low-pressure steam system lent itself to a process termed *condensing*, which allowed for exhaust steam from the engines to be recycled into water and returned to the boilers via a small reservoir and filter called the *hot well*. Recycling water was an important function on oceangoing steamships, which did not have a constant supply of freshwater for the boilers, since ocean water contains too many salts and dissolved minerals for safe usage. Therefore, a condenser was often a necessary large piece of equipment that took up a great deal of space in the engine room. Condensers were rare on high-pressure riverboats, mostly because a constant supply of fresh river water could be provided for the boilers.

The most common type of condenser up to the 1870s was the jet. It worked by mixing a jet of cold water with the exhaust steam, condensing the steam and creating a vacuum, which allowed for pumping into the hot well. Jet condensation had no real competition until the 1830s, with the

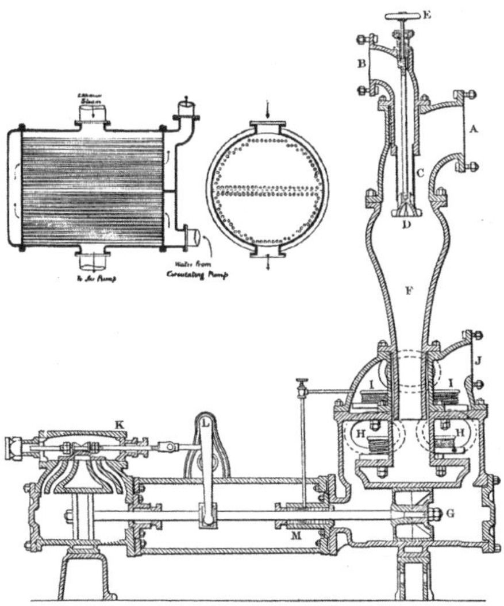

Figure 1.10. *Upper left*: Cross-section of a typical surface, or closed, condenser. As labeled on the drawing, exhaust steam from the engine enters at the top and is forced to pass downward through a bank of cold water tubes supplied with water from a circulating pump. The movement of the water through the tubes is marked by the small arrows. The steam condenses, and any stray oil and air are drawn out at the bottom by the air pump. *Lower right*: Cross-section of a typical jet condenser with accompanying pump. *A* is the passage for exhaust steam from the engines, *B* is the passage for cold water, *C* is the water jet tube, *D* is the jet nozzle, *E* is the control valve for the nozzle, and *F* is the chamber where the steam and the water mix, condensing the steam. The condensate is then moved by pump *G* up and out through outlet *J*. *K*, *L*, and *M* are the valve mechanism, linkages, and shaft packing, respectively, for the small steam engine at the left, which drives the pump.

development of the closed (or surface) condenser by Samuel Hall.[52] Surface condensing worked as a sort of reverse boiler: cold water tubes were surrounded by exhaust steam. As the steam cooled, it created a partial vacuum that allowed for pumping and the removal of air by the air pumps attached to the condenser.[53] Both condenser types (fig. 1.10) were considerably less common on American rivers, but they were used when there was an assurance of deep, relatively clean water. The surface condenser came to be the preferred type.

Other auxiliary equipment was common in both riverboats and ocean vessels. By the Civil War era, systems had been introduced for powering capstans, fan blowers, bilge pumps, firefighting equipment, and boiler feedwater devices. The steam whistle was also introduced on rivers in the 1840s, but it is not known to have been used on any Confederate ironclad. The other systems were common by the 1860s, but no one ship necessarily had all of them. The fan blower in particular was a recent innovation. In this system a small vertical boiler, usually called a *donkey*, coupled to a small engine turned a fan or fans placed in the smokebox, the entrance to the smokestack. This created a forced draft for the boiler fires,

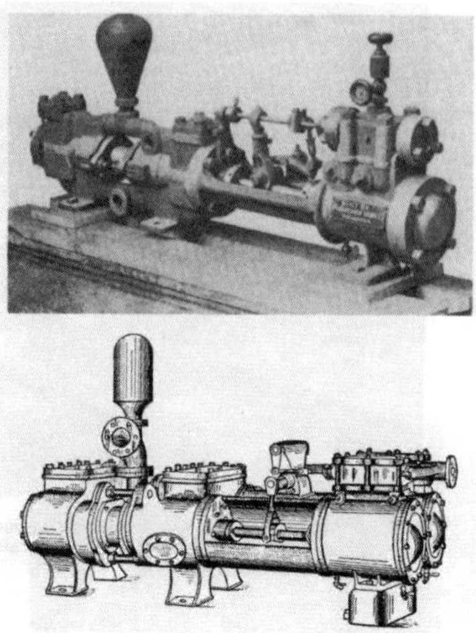

Figure 1.11. *Top*: Photograph of Worthington simplex (single-cylinder) pump. *Bottom*: Drawing of Worthington duplex (two-cylinder) pump. In both views, the small steam engines operating the device are at the right. They are connected directly to the pump at the other end by the shared piston rod between them. The balloon-shaped extensions are air expansion chambers.

allowing them a better supply of air and therefore better combustion of fuel. Despite these benefits, only five homebuilt Confederate ironclads—the *Tennessee,* the *Tuscaloosa,* the *Huntsville,* the *Missouri,* and the uncompleted *Mississippi*—are known to have used this system. Forced draft fans never became common during the Civil War because of the scarcity of materials for them, the excessive wear caused by soot and grit, and the unsubstantiated but common belief that a natural draft was better.

Donkey boilers and their engines could be used to power other auxiliaries such as steam capstans, employed to take in or let out an anchor chain or heavy lines. This arrangement was not common on warships because of their abundance of manpower, but at least one ironclad, CSS *Missouri,* had it. Steam-driven capstans were considerably more abundant on sailing merchantmen, which had small crews for economical reasons.

The rest of the auxiliary machinery carried in steamships during the 1860s related to boiler feedwater and pumping. Pumping duties not directly connected with the boilers were absorbed by the hand-powered main force pump. Two to four men, with one man or a pair of men on each end gripping the handles of the seesawlike pump lever, were normally required to operate it. Every vessel had at least one force pump placed near the boilers. In the early days, before steam pumps were common on ships, the main force pump was used for filling (feeding) the boilers, making hydrostatic tests, and emptying the bilges. This was time-consuming and

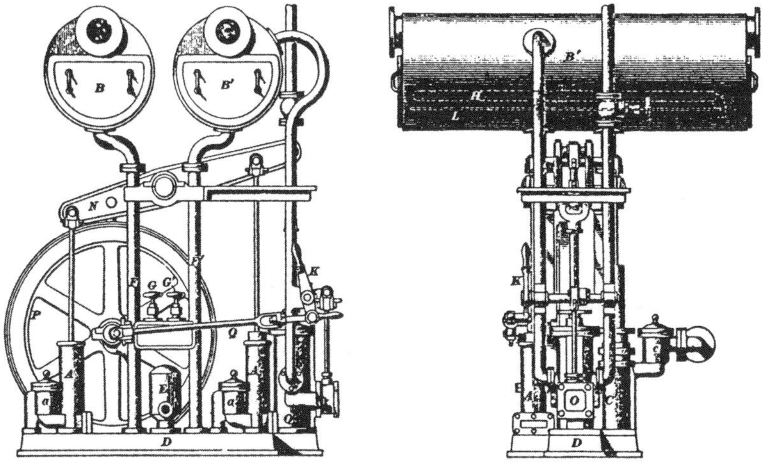

Figure 1.12. Typical doctor engine with two feedwater heaters. In the profile view on the left, *A* and *A'* are the water pumps, *a* and *a'* are air expansion chambers, *B* and *B'* are the feedwater heater access plates, *D* is the bedplate, *E* is an auxiliary attachment point, *F* and *F'* are the cold water drain pipes from the feedwater heaters, *G* and *G'* are control valves, *O* is the engine cylinder, *K* is the reversing lever, *Q* is the reversing gear link, *P* is the flywheel, and *N* is the engine beam.

strenuous work, but even when more automated systems came along the manual pumps were retained as backup.

Vessels were fitted with improved and powerful pumps driven by steam beginning in the 1860s. Probably the best of these was the type invented by Henry Worthington (fig. 1.11). One design was a compact single-cylinder, or simplex, pump, in which the steam and water pistons were attached to the same rod.[54] The other design was a larger two-cylinder, or duplex, pump and was widely used on larger vessels.

The final sets of auxiliary systems in common use by the time of the Civil War were related to boiler feedwater—the most vital system in many ways, for with too little water a disastrous explosion could occur, and with no water no steam could be produced. To greatly reduce either possibility, steamboats were required to have more than one feedwater system.[55]

The earliest steamships had pumps that ran directly off the main propelling engine. Although these pumps were simple and effective when the vessel was in operation, no water could be supplied when the engines were stopped. This caused many boiler explosions. Beginning in the late 1830s, something called the *doctor engine* became common. It got its name because it supposedly cured the ills of the previous system. A small beam engine of one or two cylinders was attached to pumps that were capable of supplying boiler or firefighting water and pumping the bilges (fig. 1.12). In this setup the piston rod of a vertically placed cylinder raised and lowered one end of a center-pivoted beam on a frame. The other end

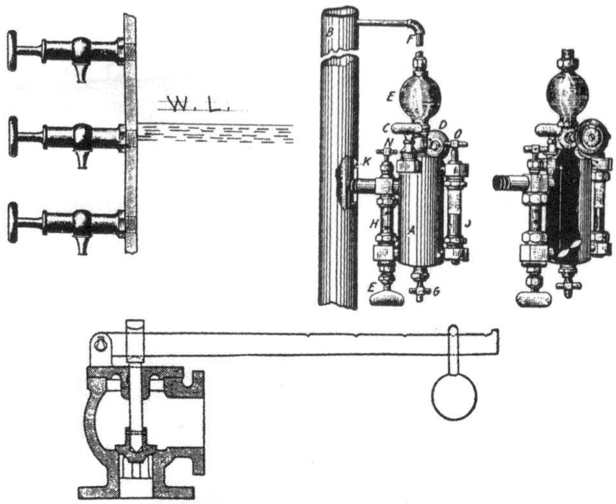

Figure 1.13. *Upper left*: Typical boiler gauge cocks. They are mounted on the boiler face where the firemen can easily see them. The ideal water level is shown. If the water level is too high, all three cocks will produce water. If the water level is too low, only the bottom cock (or, worse, none of the cocks) will produce water. *Upper right*: Typical hydrostatic lubricator with cutaway view. All hydrostatic lubricators work on the simple principle that oil is lighter than water; a small amount of steam from line *B* enters line *F*, where it is condensed in chamber *E*. Opening steam valve *D* allows the water to enter reservoir *A* (filled with lube oil), displacing the oil to the top. An internal downpipe, with its mouth even with the water surface, captures the oil and sends it up in drops through sight glass *H*. Adjusting feed valve *E* allows for increased or decreased oil amounts entering steam line *B* through pipe *K*. *N* and *O* control water admittance to the sight glasses, *G* is a drain valve, *C* is the oil fill cap, and *J* is the sight glass for determining water level in reservoir *A*. *Bottom*: Cross-section of typical lever safety valve, showing movable weight, or pea. If the steam pressure becomes too high, steam will enter the valve from the bottom, pushing up on the weight mechanism. The pressure at which the valve pops can therefore be adjusted by moving the weight on the lever.

of the beam moved the pump rods as well as a connecting rod turning the flywheel, ensuring smooth operation. Doctor engines were equipped with at least one feedwater heater that used recirculated steam to preheat the water before returning it to the boilers, lessening the chances of cold shock and explosion.

An important piece of auxiliary equipment connected with the boilers was the live steam injector, first invented by Henri Giffard in the 1850s.[56] This elegantly simple device augmented the doctor engine as a feedwater supply, contained no moving parts, and was compact and cheap. Injectors worked by firing a steam jet through a Venturi tube in the feedwater line. Sufficient velocity was therefore created to drive the mix into the boiler via a jetlike nozzle. The water-steam mixture was effectively introduced at high pressure and a high temperature like that inside the boiler.

Aside from other small components, such as gauge cocks for determining the boiler water level, simple lever safety valves, and hydrostatic lubricators for supplying cylinder oil (fig. 1.13), auxiliary systems on 1850s and 1860s steam vessels were primarily concerned with pumping and feedwater supply tasks. This was also the case on ironclad warships. Even deep down inside those formidable modern naval weapons, auxiliary systems continued to perform their mundane but important roles as indispensable components of a steam plant.

Marine Steam Engine Operations in the Mid-Nineteenth Century

Marine steam engines of the Civil War era were quite different from the complicated triple and quadruple expansion behemoths that powered most ships of the late nineteenth to mid-twentieth centuries. Steam power in the 1860s was still a relatively new technology, and as such it remained full of quirks not yet studied exhaustively. Operating a steam engine on a vessel, especially on a poorly ventilated warship such as an ironclad, required a special type of seaman, the naval engineer.

By the time of the Civil War, marine engineering was still a comparatively new field, and an engineer's status was not yet clearly defined.[57] Engineering duties encompassed a wide range of functions, and more were added as steam plants grew in complexity. The engineers of the Confederate navy were not immune to this trend, although engines in Confederate vessels were often simpler than those in other navies.

The steam engineers of the Confederate navy were organized similarly to those of the US Navy. Engineer in Chief William P. Williamson answered directly to Secretary Stephen Mallory, and under Williamson were four ranks of engineering officers. One of these was commissioned (the chief engineer), and three were warrant: the first, second, and third assistant engineers.[58] In addition to the officers, the engine and boiler room crews contained a variable number of coal heavers (these were the days before the term *stoker* came into common usage) and firemen. Coal heavers were mostly unskilled, but firemen were supposed to manage fires with different kinds of fuel and be capable of handling the engines without supervision.[59] Firemen were also required to have some skill in repair and maintenance.

Besides ensuring the smooth operation of a warship's steam plant, the engine room crew, which usually numbered about twenty on the ironclads, was required to affect the necessary repairs on board.[60] Only the most difficult jobs, such as replacing cylinders, valves, or shafts, were given to outside works such as the Charlotte Navy Yard. During repair procedures or in anything to do with the propulsion systems, the chief engineer often had authority rivaling the captain's, and many disputes among the executive officers of the vessel and the engineers arose as a result.[61]

Like the technology of steam propulsion itself, the scope of authority of naval engineering officers was not fully defined in the 1860s.

Terminology was also a bit different from that of later times. It appears that a good deal of modern confusion over the true number of engines on the ironclads and other vessels has come from the nineteenth-century use of the term *engine*. Contemporary usage defined each cylinder as an engine; therefore multicylinder machinery took the plural, *engines*. For example, a vessel fitted with two engines of 100 horsepower (HP) each would today be described as a single two-cylinder engine of 200 HP.[62]

The number of engines on the Confederate ironclads in particular has thus often been confusing. Part of the problem lies in the fact that vessels with multiple screws often could not operate the engines separately, since all units were linked. There were a few exceptions, but usually all boilers gave steam to all engines, making it impossible to run with one engine crippled unless the engines were stopped and the disabled unit disconnected.[63] Many of the twin-screw Confederate ironclads, such as the *Albemarle*, followed this pattern.

The types of fuel used by mid-nineteenth-century vessels also varied. Western river steamboats nearly always used wood, whereas naval vessels burned coal of various types—oil had not yet been developed as a fuel source for marine boilers. The Confederate navy often had to rely on a combination of wood and coal for its vessels, even the ironclads, because of its chronic shortage of resources during the war. Coal from North Carolina, Alabama, and mines near Richmond was usually fairly abundant, but an ongoing lack of available transportation hampered its delivery to the shipbuilding centers.[64] In addition, the fuel from North Carolina was considered almost as bad as wood for the amount of smoke it gave off.[65] Wood often had to be resorted to while in port and even occasionally while in action.

Starting a steam engine in the days before separate starting engines were widely utilized was a monstrous task. Once a suitable amount of coal or wood had been procured, steam was accordingly built up in the boilers. With a sufficient head of steam raised, the entire engine room crew on duty had to work together. First, all parts of the engine were slowly warmed up by live steam, then steam was bypassed from the valve chest into the condenser to expel the air, creating a vacuum and freeing up the pistons to be pushed by steam alone. When all was ready, the crew barred, or turned, the crankshaft around a few times by main force. This created a vacuum in the cylinders for the steam to act effectively on the pistons when admitted. On horizontal and vertical engines it was also helpful to ensure that the crankshafts were off dead center. Then the condenser injection valve was opened, and the chief engineer, praying for a good vacuum, moved the steam throttle up a notch or two and hoped for the best.[66]

Starting the high-pressure engines of western riverboats was a slightly different procedure. It utilized an engine's own valve gear and linkages, termed *levers* and *wipers,* as described in the previous figure. The typical setup was first configured by lifting the levers off the wipers. Next the reversing link was set to the desired direction, the levers lowered to the wipers again, and the throttle opened.[67] This procedure was made less arduous by the smaller size and greater ease of access of riverboat engines.

Once the vessel's power plant was running, typical operational procedures involved maintaining the proper boiler water level, lubricating the moving parts, monitoring the engine functions, awaiting orders from the helm, and simply trying to survive in the extreme heat. Making sure a boiler contained enough water was perhaps the most important of these tasks, since an overheated or unclean boiler had a nasty tendency to explode—especially the high-pressure systems on western river steamboats. Consequently, during each voyage, the mud drum, which collected sediment from the boiler, was blown out, or emptied, several times.[68] Although this practice was most common on river steamers, it was also necessary for coastal and harbor defense vessels like the majority of Confederate ironclads.

Monitoring engine functions in the 1860s still mostly involved simple observation and an engineer's feel for his engines. Glass gauges for monitoring steam pressure, lubricant level, and other functions were fairly uncommon, being considered unreliable, dangerous, and, at best, a "dubious innovation."[69] Safety valves were rarely equipped with any tamper-proof mechanism, and direct observation of a troublesome area was often hindered by cramped machinery space. As a result, engineers of the Civil War navies most often had to rely on sheer familiarity with their charges.

Orders to the engine room in Confederate ironclads or other mid-nineteenth-century vessels were relayed by a combination of bells and shouted orders. The pilothouses and helm stations of several ironclads were positioned directly over their engine rooms, making the dispersal of orders simpler and faster. Changing direction, or reversing, as it was usually called, was the most commonly ordered task, but as with starting the engine, it was labor-intensive. There were several different reversing gear types in common use by the Civil War era, but all functioned similarly, differing mainly in complexity. Perhaps the simplest, most common, and most versatile was the Stephenson valve gear.

The Stephenson valve gear, though simple, functioned through

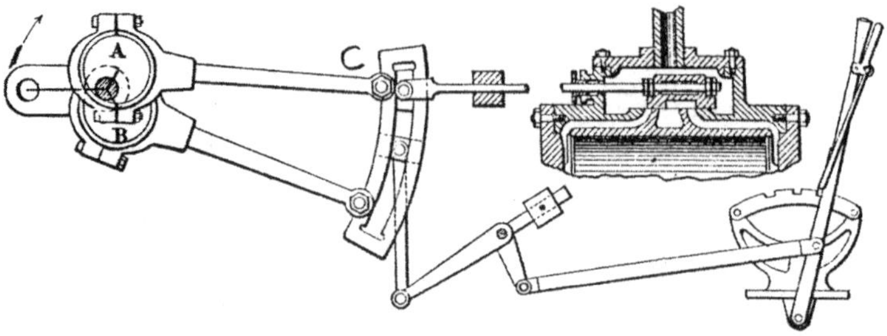

Figure 1.14. Stephenson valve gear. This setup was also commonly employed by steam locomotives, for which it was originally invented. Moving the lever attached to the link motion allows for forward, reverse, and a nearly infinite range of adjustments if the engineer's strength holds out. Eccentrics *A* and *B* are positioned at different points of rotation on the main crankshaft. *C* is the sliding link block that engages the eccentrics through the valve rod from the engine cylinder, a partial cross-section of which is shown to the right of the link block. The whole mechanism is adjusted by the control lever at the right. The three notches in the control mechanism represent forward, neutral, and reverse. Since marine engines nearly always turned propellers clockwise (represented here at the extreme left by the arrow), this engine is set at forward.

the clever use of geometric and rotational principles, which is easier to observe than describe (fig. 1.14). The main control lever (provided with three main direction points: forward, neutral, and reverse) was connected by a reach rod and a series of arms to a link that was given motion directly from the main valve rods. The motion was transmitted through the link block by a rod to an eccentric positioned on the main crankshaft. (Eccentrics are another part of the valve gear, connecting the valve linkages to the propeller shaft; they are off-center discs coupled to the main shaft that change its rotary motion into oscillating for operating the valve gear.) Different eccentrics were used for forward and reverse running. Therefore, adjusting the position of the link block determined which eccentric was put in a direct line with the valve rod. Because the eccentrics were carefully fixed at different points of rotation around the crankshaft, the position of the link ultimately set the direction of the engine.

In addition to other miscellaneous duties—such as occasionally running bilge pumps, testing fire pumps, and taking on fresh boiler water (using the so called doctor engine, if present)—a Civil War engineer's main concern was trying to endure the extreme heat of the engines and the boiler furnaces. Rarely, if ever, were provisions made in the early days of steam for proper ventilation of the belowdecks spaces other than hatches and gunports.[70] The problem was made even worse on the ironclads with their layers of insulating armor and dearth of hatches. Consequently, the amount of air reaching the furnaces was usually insufficient for ideal combustion, and this, combined with a reliance on a natural draft and an almost complete lack of insulation on the boilers and steam lines, led to excruciating conditions in the machinery spaces. It was not uncommon for the men to endure temperatures over 120 degrees Fahrenheit, and in at least one ironclad, CSS *Richmond*, the temperature was reported to occasionally exceed 150 degrees.[71] Exposed parts of the boilers, such as the uptakes to the smokestack, or breechings, could be seen to glow red even in daylight, and gases from the furnaces often permeated the living quarters, further contributing to the ill effects of serving on an ironclad for any length of time.[72]

There is little wonder that it was typical for an engineer's appearance to vary greatly between watches. He went on duty neatly attired in a clean uniform, groomed, and full of energy. Coming off his watch, he was usually soaked in sweat and condensed steam; covered in grease, oil, or coal dust; and worn out from exertion and heat exhaustion.

In short, the duties involved in maintaining and operating an 1860s marine steam plant were varied, dangerous, and labor-intensive. Engineers were required not only to be familiar with a high degree of technical detail but also to operate from gut instinct and sheer strength. Their relatively new place in the hierarchy of naval officers often brought them into conflict and led to confusion about the full extent of their duties. During the Civil War, these problems remained common, but as with many other aspects of steam engineering at that time, progress was made.

The veteran engineers of the Civil War navies observed and experienced firsthand the transition from combined sail and steam power to vessels driven purely by steam. This change happened especially early on the ironclads. Experience gained during wartime service allowed for the effective training of the next generation to serve on vessels built after the transition from sail to steam and from wood to iron was complete. Similarly, in their time the marine steam engineers of the Civil War were trained by men who had seen the very earliest marine steam engines in operation, one of the first great revolutions of the nineteenth century.

The next revolution in American warship design began with the conversion of a few unlikely vessels into the first Civil War ironclads.

TWO

CONVERSIONS

To what strange uses things are sometimes put.
—Chief Engineer H. Ashton Ramsay, Confederate States Navy

The vessels converted into ironclads represented the rush of the Confederacy in the early days of the Civil War to prepare what it considered the ultimate weapon against the US Navy. The *Manassas*, the *Virginia*, and the *Baltic* had been built for greatly different purposes in their previous incarnations, and each was selected for conversion based on a set of characteristics thought desirable for an ironclad. Each vessel in effect became a prototype, but ultimately the *Virginia* emerged as the best: the relatively light damage received from USS *Monitor* at the Battle of Hampton Roads justified the simple to construct, angled shield concept, and the structure became a distinguishing feature of Confederate ironclads.[1]

The *Virginia* became by far the most famous Confederate ironclad, but the other two conversions remain fairly obscure. The *Manassas*, at least, played a prominent role in the war's first major turning point: the determined but unsuccessful Confederate defense of New Orleans and the Mississippi.

CSS *Manassas (Ex–Enoch Train)*

CSS *Manassas* was the first armored ship to serve in the Civil War, but its career began very differently from as the Boston Harbor towboat *Enoch Train*, launched in December 1854 (fig. 2.1).[2] The vessel's hull was completed by the famous clipper-ship building company of J. O. Curtis at its yard in Medford, Massachusetts. The hull was designed by Samuel Hartt Pook and strongly built with fine lines.[3] The completed *Enoch Train* had a registered tonnage of 385, a length between perpendiculars of 128 feet, and a maximum beam of 28 feet.[4] With the woodwork complete, the vessel required suitably powerful and well-built steam machinery. Harrison

Figure 2.1. This painting by maritime artist William Pollard (1801–1872) represents the *Enoch Train* as completed. The hull form and pilothouse forward clearly denote this vessel as a towboat. The fact that it is a twin screw is revealed by the exposed tip of a propeller blade beneath the American flag aft. The strange device on the aft pilothouse top is presumably the galley stovepipe, whereas the steam whistle can be seen on the forward base of the smokestack. The steam escape pipe and stokehold ventilators are behind the stack. The tall masts, rigged for auxiliary sails, were common to early towboats.

Loring's famous City Point Works in South Boston was the firm chosen for its manufacture.[5]

City Point Works had achieved a national reputation for excellence in steam machinery by the 1860s. Harrison Loring had established a machine and boiler shop in South Boston in 1847, and his business grew and prospered. Ten years later Loring purchased land along the waterfront for a combined shipyard and machinery firm. This became City Point Works, "one of the first shipyards in the United States devoted to the construction of iron steamships, engines, and boilers."[6] Loring's skilled establishment turned out several merchant ships before the Civil War and produced machinery for a number of other ships. During the war, two monitors and three sloops of war were constructed there. Admiral David G. Farragut's famous flagship USS *Hartford* also received its machinery from City Point Works.

The machinery built by City Point Works was highly praised.[7] This was actually a major reason for the *Enoch Train*'s eventual conversion into an ironclad. The powerful towboat was provided with independent twin-screw propulsion. During the metamorphosis into the *Manassas* it was thought that this would give more than enough power to function as a ram. As usual, the sheer burden of the protective wood, armor, and weaponry placed on the hull was underestimated. Those problems, however, were unanticipated when Loring built the *Enoch Train*'s power plant.

A fair amount of detail exists regarding the machinery arrangement and specifications, although no technical drawings of the ship are

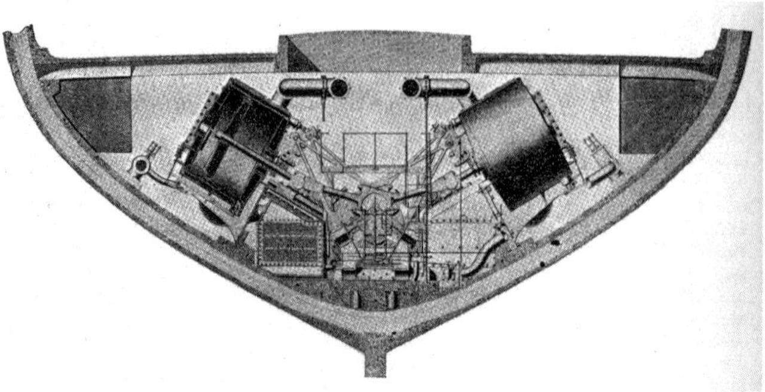

Figure 2.2. Inclined inverted direct-acting engines of the 1863 screw frigate HMS *Constance*. Each cylinder is placed at an angle above the turn of the bilge with its piston and connecting rods transmitting power directly to the propeller shaft at the center. Although this particular example is a British compound, its layout is representative of regular inclined engines as installed on CSS *Manassas*, the ex–*Enoch Train*. In that case, each engine cylinder turned its own propeller shaft.

known. The towboat's powerful and unusual (for the 1850s) twin-screw propulsion system drew its share of comments, especially during an 1883 Supreme Court case involving patent infringement on dredging methods. One of the *Enoch Train*'s former operators provided details on the machinery during the hearings.[8]

Two single-cylinder inclined direct-acting engines, each turning one three-bladed screw, drove the vessel; its top speed was about 15 knots. The engines were low pressure and, like most low-pressure systems, were equipped with condensers for recycling spent steam back into the boiler water. The engine cylinder diameter was 36 inches, the stroke was 32 inches, and the propellers were about 9 feet in diameter. The HP produced was about 450. The engines could be worked independently, providing improved maneuverability and redundancy if one engine became disabled. The *Enoch Train* also was fitted with one doctor engine that powered a large wrecking pump capable of throwing 15,000 gallons per hour.[9] All the machinery was powered by one or two low-pressure boilers of uncertain type and size.[10]

The *Enoch Train*'s inclined engines were a variant of the common direct-acting form. This particular layout was useful for small vessels with multiple screws. On the *Enoch Train* each engine was most likely mounted above an angled outboard of each propeller shaft. Such an arrangement was typical on other vessels (fig. 2.2). No contemporary sources record anything unusual about the machinery.

No information has been found on the *Enoch Train*'s first four years of service. During that time the boat was probably performing the duties it was intended for and staying out of the spotlight of history.

In 1859 the *Enoch Train* was brought to New Orleans for towing and dredging work. Its powerful twin-screw propulsion system was ideal for operations on the Mississippi River and in May 1861 led a group of New Orleans businessmen to purchase the vessel for conversion into an ironclad ram. Steamboat Captain John A. Stephenson supervised the *Enoch Train*'s conversion at the Algiers, Louisiana, shipyard of John Hughes & Company.[11]

Stephenson was a prominent New Orleans citizen and the secretary of the New Orleans Pilots' Benevolent Association. His reason for converting the *Enoch Train* was simple: privateering money. The Confederate government encouraged privateers—especially in the Civil War's early days, when a navy had not yet been organized—with easily obtainable letters of marque and a large bounty (20 percent of a prize's value).[12] In May 1861 Stephenson, possibly with foresight about fighting federal blockaders, lobbied several officials, including Jefferson Davis, with a proposal to "alter and adapt some of our heavy and powerful towboats on the Mississippi as to make them comparatively safe against the heaviest guns afloat."[13] The proposal also outlined the advantages of reinforcing the bows of such boats for rams. This was the genesis of the *Enoch Train*'s conversion and singular design as the *Manassas*.

Captain Stephenson and his partners in the privateering venture had little difficulty obtaining support. Local speculators were apparently either very patriotic or impressed with the venture as a business opportunity—fifty thousand dollars was raised the first day.[14]

The *Enoch Train* was the obvious choice for conversion because of its strong bow and twin-screw propulsion. It was moved across the river into one of the eight drydocks at Algiers, and work began; the new name *Manassas* was chosen for the recent Confederate victory at the village of the same name in Virginia. The boat was ultimately razed to main deck level and given a convex wooden shield covered with railroad iron; only one forward firing gun was placed within it. The bow was also lengthened and restructured into a ram, but its exact form is unknown (fig. 2.3). Unfortunately, the addition of so much extra weight reduced the *Manassas*'s speed considerably; 8 knots were obtainable only when steaming downstream with the river current.[15]

When the *Manassas* was completed, its measurements were a total length of 143 feet, a 33-foot maximum beam, and an estimated 10.5-foot draft.[16] It is unknown if modifications were made to the *Enoch Train*'s engines during the conversion into the *Manassas*, but it appears that the low-pressure boilers were replaced with a set of western river–type high-pressure flue boilers.[17] Since the *Manassas* was intended as a ram, greater power generation for greater speed was required, and high-pressure boilers could provide that in a smaller space than the original boilers,

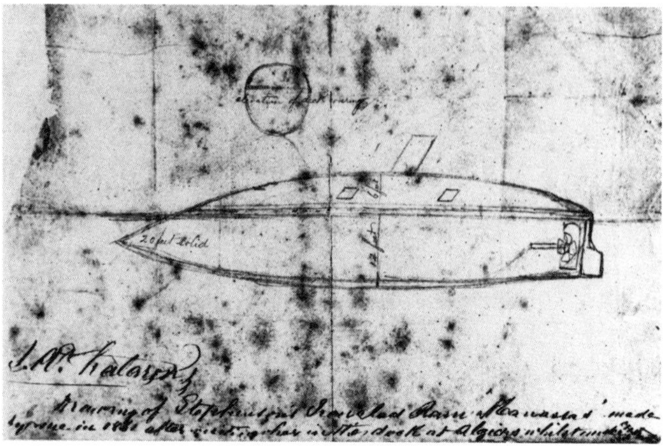

Figure 2.3. Rough sketch of the *Manassas* by J. A. Chalaron. The handwriting at the bottom, below Chalaron's signature, reads, "Drawing of Stephenson's [sic] Ironclad Ram 'Manassas' made by me in 1861 after visiting her in the dock at Algiers whilst under construction." This is probably the best-known contemporary image of the *Manassas*, but most of the details shown are unclear or do not quite match other firsthand accounts.

which were probably quite worn by 1861. In addition, this could account for the two smokestacks shown in contemporary images of the *Manassas*, since a set of three or four flue boilers commonly warranted two sets of uptakes.

The *Manassas* remains unique as the only ironclad ever intended for privateering, but it never pursued that line of work. Launched in mid-August with a letter of marque applied for and a crew enlisted, the new ironclad was seized in September 1861 by Confederate navy forces in an attempt to bolster the defense of New Orleans.[18] The seizure of the vessel by professional sailors under Lieutenant Alexander F. Warley was an interesting and amusing incident that in some respects demonstrated the antipathy of the state and private naval forces toward being taken over and incorporated into the new Confederate States Navy. Midshipman James Morris Morgan, a participant in the commandeering, wrote the following:

> On arriving alongside the ram [in a boat from the warship CSS *McRae*] we found her crew lined up on the turtleback, swearing that they would kill the first man who attempted to board her. There was a ladder reaching to the water from the top of her armor to the water line. Lieutenant Warley, pistol in hand, ordered me to keep the men in the boat until he gave the order for them to join him. Running up the ladder, his face set in grim determination, he caused a sudden panic among the heroic crew of longshoremen who incontinently took to their heels and like so many prairie dogs disappeared down their hole of a hatchway with Mr. Warley after them. He drove them back on

> deck and then drove them ashore, some of them jumping overboard and swimming for it.[19]

Thus the *Manassas* became a Confederate States Ship (CSS) and at the same time acquired a bold new commander. Warley's bravado and courage were typical of the man who would soon lead the first American ironclad into battle.

By 1861 Alex Warley was one of the navy's most competent and experienced junior officers, but as a lieutenant he was also just senior enough for the opportunities that opened in the Confederate navy.[20] Warley was born in Walterboro, South Carolina, on July 29, 1823, and joined the navy in 1840 as an acting midshipman. Even from a young age he was marked by his intelligence, boldness, and temper. The latter trait got him in trouble (four courts-martial), but it was mostly because of a glut of senior officers filling the ranks that it took him fifteen years to reach lieutenant. He served in a wide range of duties on several different ships (on both coasts of North and South America, in peacetime and during the Mexican War) and was also briefly stationed ashore. Warley sailed in the famous paddlewheeler USS *Mississippi* to Singapore and Hong Kong at the height of the Second Opium War (1856–1860) and returned to the United States in 1860, just before the Secession Crisis.[21]

When Warley returned from his Asiatic cruise in January 1860, the great rift between the Northern and Southern states had begun. Even before arriving home Warley was aware of the situation through both his brother Felix, who left his law practice to join the artillery militia, and his influential cousin and former congressman William P. Miles. Warley fully intended to resign his commission in the US Navy at the first opportunity and join the South Carolina naval forces. He did just that, then headed to Charleston to command a small battery of Dahlgren guns on Morris Island during the momentous days leading up to and immediately following the bombardment of Fort Sumter. Warley was then ordered to New Orleans, where he served as a first lieutenant on CSS *McRae* and was later appointed by Captain George N. Hollins (in charge of New Orleans naval defenses) to commandeer and captain the private ironclad *Manassas*.[22]

After his courageous capture of the ram, Warley was able to retain only three men: First Mate Charles Austin, Chief Engineer William H. Hardy, and local pilot J. Stevens Mason.[23] The rest of the crew was formed from volunteers. Warley was displeased with his first command: he described the *Manassas* as a "bug-bear . . . no power, no speed, no strength of resistance and no armament."[24] In reality the little ship was armed with one forward firing gun, but its primary weapon was its ram. Unfortunately, the *Manassas*'s speed for ramming was severely limited by its heavy armor, and the conditions in the machinery space were nearly intolerable: the temperature was reported at over 130 degrees Fahrenheit. This was also a problem in many other ironclads; being the first ironclad, the *Manassas* simply set the standard.

Warley was forced to take his new charge into action almost immediately at the Battle of the Head of Passes on October 12, 1861. No firsthand engineering accounts of this battle regarding the *Manassas* are known to exist, but the event was vividly recounted by multiple sources as the first wartime use of an ironclad ram in combat.

By the end of September 1861 a fleet of four Union vessels (the steam sloop *Richmond*, the sailing sloops *Vincennes* and *Preble*, and the sidewheeler *Water Witch*) had occupied the strategic Head of Passes, the site about 15 miles above the mouth of the Mississippi where the river divides into several channels. New Orleans and the river above that point were thus sealed off to blockade-runners.[25] In order to reopen this vital stretch, the *Manassas*, leading a little Confederate fleet (the gunboats *Calhoun*, *McRae*, *Jackson*, *Tuscarora*, *Pickens*, and *Ivy* as well as tugboats towing fire rafts) headed downriver on the dark night of October 12 to attack the federal fleet. Reaching the scene, Warley steered the *Manassas* toward USS *Richmond*, which he apparently did not recognize as the ship he had left ten months earlier.[26]

"Let her out, Hardy! Let her out now," Warley roared to his chief engineer, and the firemen piled tar, tallow, and sulfur into the fires as the *Manassas* shuddered along at ramming speed. The ram crashed into the *Richmond*'s side, stoving in that vessel's planks about 2 feet below the waterline, but the *Manassas* itself suffered an upending of one of the engine mounts, putting one screw out of commission. At the same time, the stacks were knocked over from flying shells and the hawser of a coaling schooner alongside the *Richmond*. The *Manassas* actually suffered more damage than its target vessel did. After signaling for the fire rafts to be brought downstream, Warley made the decision to withdraw.[27]

The federal fleet was thrown into chaos after the *Manassas* rammed the *Richmond*. Broadsides were fired in every direction, and long-range exchanges with the other Confederate vessels lasted through the night. The Confederate fire rafts caused the blockaders to scatter downriver, where they ran aground and generally did not acquit themselves well; despite these difficulties, no major damage was sustained. Within days the blockade was reinstated at the river mouth. In the meantime Warley grounded the *Manassas* to prevent further damage, and the ship had to be towed back to New Orleans. He was hailed as a hero, and his pioneer vessel was put into drydock for extensive repairs.[28]

The *Manassas*'s performance at the Head of Passes revealed that the ram's abilities were limited at best—it was more of a psychological weapon than anything else. It was also revealed that the bracing hitherto provided for the machinery was insufficient for ramming, and this was updated accordingly. Once the repairs were completed, Warley and his charge patrolled the Mississippi River as far north as Memphis, but they discovered that the *Manassas* had great trouble effectually stemming the current. Warley returned to New Orleans in March 1862. By that time Flag Officer David G. Farragut and his Union forces were massing in the Gulf of Mexico in preparation to seize New Orleans and the lower Mississippi River once and for all.[29]

Upon news of the approach of Farragut's fleet, all available Confederate units were called to prevent any enemy vessel from passing upriver. The stand was made at the site of two large forts, Jackson and St. Philip, directly across the river from each other about 60 miles below New Orleans. If the federal fleet could be delayed there by the Confederate vessels and river obstructions, the forts would, in theory, be able to repulse it.

Lieutenant Warley wrote about this second and final fight of his ship, the Battle of Forts Jackson and St. Philip (April 24, 1862). His account is brief but informative. It not only relates the *Manassas*'s maneuvers and problems during the battle but also offers a glimpse into what it must have been like to serve on such a warship during the Union's climactic bid to approach and take New Orleans. All the Confederate naval units were involved, but the *Manassas* played a prominent role because Warley threw everything he had at the federal fleet. On the performance of the *Manassas* during one of its ramming runs, Warley wrote, "The *Manassas* was driven at her [USS *Brooklyn*] with everything open, resin being piled into the furnaces. The gun was discharged when close on board. We struck her fairly amidship; the gun recoiled and turned over and remained there, the boiler started, slightly jamming the Chief Engineer, Dearning, but settled back as the vessel backed off."[30]

Once again the *Manassas*'s machinery was coming off its bedplates from the shock of heavy ramming. The ironclad had already hit USS *Mississippi* (another ship on which Warley had sailed) on the way down to meet the rest of the Union fleet. Unfortunately for the Confederates, neither ramming event did enough damage to sink the target vessel.

After the nearly disastrous encounter with the *Brooklyn*, Warley took his battered ship back upstream, witnessing the utter defeat of the Confederate fleet on the way. Farragut did not stop to engage the Confederate warships and forts as expected; instead he proceeded directly to New Orleans at any cost. The powerful Union warships thus engaged the outnumbered and disorganized Confederate fleet almost at their leisure, although the fight was a close and bloody one. The *Manassas*, in the forefront of most of the action, took a tremendous beating. Warley wrote, "Steaming slowly up the river—very slow was our best. . . . The *Manassas* was helpless. She had nothing to fight with, and no speed to run with. I ordered her to be run into the bank on the Fort St. Philip side, her delivery-pipes [water intakes] to be cut [flooding the hull], and the crew to be sent into the swamp through the elongated port forward, through which the gun had been used."[31] After the abandonment, federal sailors reached the stricken hulk and set fire to it. The burning wreck dislodged from the bank and drifted downriver as several Union ships poured broadsides into the thin iron shield.[32] Finally the battered, burned, and punctured little craft blew up and sank. So ended the first American ironclad.

Alexander Warley was taken prisoner when Forts Jackson and St. Philip surrendered on April 30, 1862. He was held at Fort Warren in Boston Harbor and was released in a prisoner exchange on July 31.[33] Warley was later the executive officer of the ironclad CSS *Palmetto State* at Charleston, then the captain of its sister, the *Chicora*; his last command was CSS *Albemarle*. He will be encountered again in the subsequent chapters that discuss those particular ironclads.

CSS *Virginia (Ex–*USS *Merrimack)*

The story of CSS *Virginia*, more commonly known by its original US Navy name, USS *Merrimack*, is well told elsewhere. This most famous of Confederate ironclads has been the source of exhaustive study, and fortunately its machinery systems are well-known. Although the *Manassas* was technically the first North American ironclad to be completed, it did not make the kind of lasting impression on Confederate ironclad design that the *Virginia* did. No other homebuilt ironclad had so much hope vested in its completion for the Southern cause, and no other involved so many of the Confederacy's most prominent naval officials in its design. Long before making history as the *Virginia*, however, the ship was making history as USS *Merrimack*.

The US Navy was known for building very large, powerful frigates that defied traditional classification. USS *Constitution*, the veteran ship of the War of 1812 known as Old Ironsides, is the most famous. In April 1854 a new round of large frigate construction began with the *Merrimack* class of screw steamers (fig. 2.4). The *Merrimack* and its sister ships were very large and powerful, but their engines were not intended as their primary source of propulsion. Nevertheless, they created quite a stir in Europe.[34]

The *Merrimack* itself was the first US steam frigate, built at the Charlestown Navy Yard in Massachusetts from July 1854 to February 1856. It was also the first auxiliary steam-powered propeller-driven frigate in the world.[35] The design was copied on four sister ships (the *Wabash*,

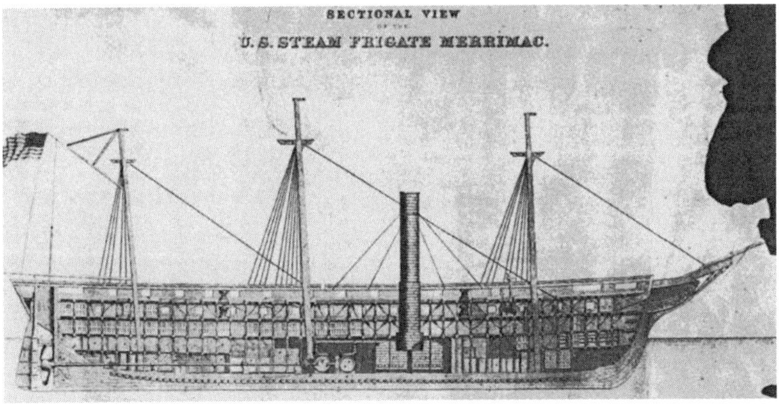

Figure 2.4. Sectional view of the *Merrimack*, showing machinery arrangement. Note the well for raising the Griffiths patent screw at the stern and the lower berth deck just above the boiler tops. This deck became the main deck of CSS *Virginia*—everything above it was burned away when the *Merrimack* was scuttled.

the *Roanoke*, the *Minnesota*, and the *Colorado*), but a different engine type was used for each. The machinery in these primarily sail-driven ships was intended to provide auxiliary power and to simplify maneuvering in and out of ports.[36]

The massive power plant installed on the *Merrimack* was built by the West Point Foundry in Cold Spring, New York, one of the largest and most experienced manufacturing facilities in the United States before the Civil War. The *Merrimack*'s engines were designed by the foundry's superintendent, Captain Robert Parker Parrott, and the boilers were designed by US Navy Engineer in Chief Daniel B. Martin. Despite the collaborative efforts of these foremost steam engineers working at one of the foremost manufacturing establishments in North America, the finished product ultimately reflected the relative youth of steam engine design.[37]

The West Point Foundry (originally called the West Point Iron and Cannon Factory) was established in 1817 by a prominent local citizen named Gouverneur Kemble, with financial backing from a group of New York businessmen (including the first graduate of West Point Military Academy, General Joseph Swift).[38] It was supervised beginning in 1829 by Captain Parrott—the inventor of the Parrott rifle, which was used extensively by both sides in the Civil War—and soon came to specialize in producing steam engines in addition to artillery.

By 1851 more than four hundred men were employed, and the sprawling complex comprised an iron foundry, a brass foundry, a pattern shop, a smith's shop, a machine shop, and a boiler shop. All the equipment necessary for producing steam machinery was present: three furnaces, one

7-ton trip hammer, two tilt hammers, twenty-eight turning lathes, and three planing machines. This was the source of the *Merrimack*'s propeller, shaft, four boilers, and engines. Such an impressive collection of quality machining tools and talent seemed to guarantee excellence, but the US maritime steam engine industry was still in its infancy in the 1850s. The *Merrimack*'s engines were soon found to have serious weaknesses. The West Point Foundry ultimately had far more success (especially during the war years) in crafting artillery pieces and shells. The foundry even received a congratulatory visit from President Abraham Lincoln in 1862.[39]

When it was completed, the *Merrimack* was powered by four boilers connected to two horizontal single-cylinder return connecting rod (or back-acting) engines turning a single screw. This machinery was only auxiliary to the frigate's 48,757 feet of canvas. The *Merrimack* could make 10.656 knots underway using both sail and engines, but only 6 knots using just engines.[40]

The *Merrimack*'s hull and machinery statistics were impressive: the ship displaced 3,211 tons fully loaded, with an overall length of 275 feet, a beam of 56.5 feet, and a draft of about 24 feet.[41] (After the ship's conversion to CSS *Virginia*, these measurements changed to 3,500 tons; an overall length of 280 feet 9 inches, a beam of 51 feet 2 inches, and a depth of hold equal to 21 feet 5 inches, resulting in a draft of about 22 feet.[42]) (*Depth of hold* was an old way of measuring a ship's interior space from the lowest part of the hull at its midpoint to the underside of the uppermost full-length deck.) The installed machinery included four Martin vertical watertube boilers, each 15 feet high, 14 feet wide, 12 feet deep, and 28 tons.[43] The watertubes were brass, each 2 inches in diameter and 39 inches long, and there were seven hundred per boiler; a strong double-riveted iron shell plate 0.375 inches thick protected each boiler assembly. The boilers operated at a maximum pressure of 16 PSI and had a total aggregate heating surface of 12,537 square feet.[44]

Each of the *Merrimack*'s boilers was equipped with one feedwater heater under the floor plates, mercurial steam pressure gauges, simple lever safety valves, and at least one glass pressure gauge. In addition, the ship was fitted with two Worthington duplex pumps, each 8 inches in diameter, used for supplying boiler feedwater, pumping the bilges, or fighting fires.

The two engines were low pressure and jet condensing, each with a cylinder diameter of 72 inches and a stroke length of 36 inches. Each cylinder was connected to the crankshaft (14.5 inches in diameter) by twin piston rods (7 inches in diameter). Each air pump was 42 inches in diameter with a 27-inch stroke, and each steam pipe was 18 inches in diameter. The engines could not be operated independently. The average operating

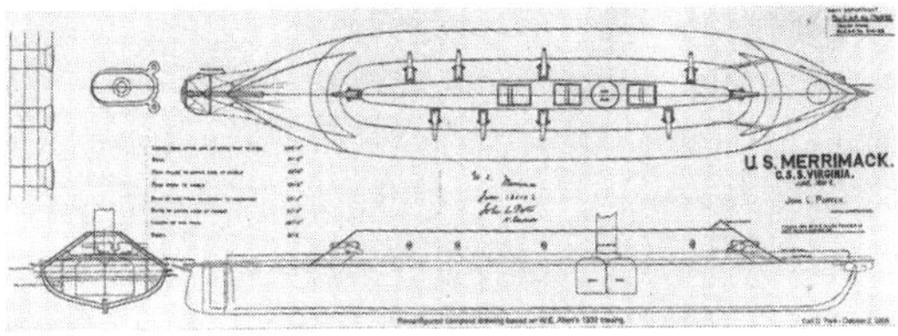

Figure 2.5. John L. Porter's plans of the *Virginia*, reconfigured and redrawn for clarity by Carl D. Park in 2005. Although the engines are not shown, the location of the boilers and their uptakes to the smokestack are clearly shown. Note also the complicated series of cables and supports for moving the rudder shown in the top view.

pressure for the *Merrimack*'s engines was 25 PSI at about 46 RPM, and reverse gearing was accomplished with Stephenson's link motion. The total gross effective HP was estimated at 869 on the propeller shaft with an estimated requirement of 103 HP just to turn the engines over (the entire engine room crew was required for starting them!). A further 65 HP was reputedly lost because of friction on the propeller shaft.[45]

The arrangement of the *Merrimack*'s propulsion system was also unique because of the unusual type of propeller employed. The single wrought-iron propeller shaft was forged in two sections, each 30 feet long and 13.5 inches in diameter, which turned a massive (17 feet 4 inches in diameter) two-bladed screw. This particular design was patented by an Englishman named Robert Griffiths in 1849 and featured adjustable blades for aid in sailing quality. It also allowed for the raising and removal of the propeller while underway with the ship under sail alone.[46]

There were serious defects that severely hampered the *Merrimack*'s performance during its five-year service before its layup for a comprehensive overhaul at the Gosport Navy Yard in Norfolk, Virginia, in February 1860.[47] The primary sources of trouble were a poor vacuum in the cylinders because of air leaks, poor arrangement of the air pumps and condensing portion of the machinery, and excessive vibration at speed.[48] It was also eventually revealed that the engines were quite underpowered for the size of the ship.[49]

The design problems and overall unsuitability of the engines were widely known before the *Merrimack*'s layup. Benjamin Franklin Isherwood (1822–1915), the engineer in chief of the US Navy from 1861 to 1869, stated in an 1863 report that "the enormous cylinder capacity caused the steam to have a low average pressure on the pistons throughout the stroke, and the vacuum, owing to air leaks, was very poor. The arrangement and proportions of the air-pumps were faulty and productive of the same results."[50] This expert assessment was later corroborated by the

Confederates who had converted the *Merrimack* into CSS *Virginia* beginning in the summer of 1861.[51]

Events in the prelude to war from 1860 to 1861 rapidly overtook the *Merrimack* as it languished at Norfolk. When the state of Virginia seceded from the Union, the security of the great naval base was much in doubt. The yard was evacuated on the night of April 20–21, 1861, in the face of an approaching Confederate force, and the federal forces hastily set fire to everything that could not be removed, including the *Merrimack*. The ship was unable to leave because the engines were disconnected and parts were missing in the first steps of a long-awaited machinery overhaul. The frigate sank at its moorings before the fire reached below the berth deck, preserving the boilers, engines, and lower hull.

The inability of the Confederacy to manufacture adequate steam machinery at the dawn of the Civil War led directly to the decision to convert the *Merrimack* into the ironclad ram CSS *Virginia* in 1862 despite the ship's previous lack of effective power and reliability. The burned-out hulk was raised and rebuilt into the prototype Confederate casemate ironclad (fig. 2.5), and the power plant was repaired as much as possible under the circumstances. Considering the inherent machinery defects and the traumas suffered from scuttling, it is remarkable that the Confederates kept the engines running as well as they did during the *Virginia*'s brief career.

During the planning process in June 1861 for building an ironclad, Confederate Navy Secretary Stephen Mallory summoned Naval Constructor John L. Porter, Lieutenant John M. Brooke, and Confederate Navy Engineer in Chief William P. Williamson to Richmond to draw up plans. According to Brooke, "We learned that there were no suitable engines in the South. Williamson then said he thought the engines of the *Merrimac* [*sic*] could be used, but that the vessel would necessarily draw as much water as the *Merrimac*, and it would not be worthwhile to build a new hull, as enough of the old hull remained to carry out the plan. Mr. Porter and I thought the draught too great, but that we could not do better. We so reported to the secretary, who concurred."[52] An official report to Mallory, dated June 25, 1861, soon followed:

> SIR: In obedience to your order, we have carefully examined and considered the various plans and propositions for constructing a shot-proof steam-battery, and respectfully report that in our opinion the steam-frigate *Merrimac*, which is in such condition from the effect of fire as to be useless for any other purpose without incurring a very heavy expense in her rebuilding, can be made an efficient vessel of that character, mounting 10 heavy guns, . . . and from the further

> consideration that we cannot procure a suitable engine and boilers for any other vessel without building them. . . . The bottom of the hull, boilers, and heavy and costly parts of the engine being but little injured, reduce the cost of construction to about one-third of the amount which would be required to construct such a vessel anew.[53]

Thus the burned out hulk of the *Merrimack* became the prototype Confederate ironclad, and its machinery was once again called on to move a ship too large for the amount of power generated.

The three men most intimately involved with the *Virginia*'s design and construction were the Confederacy's most prominent leaders in its shipbuilding, steam engineering, and naval artillery programs. Brooke gained some fame as the inventor of the eponymous rifle, the best naval ordnance piece used by the Confederacy.[54] Porter's and Williamson's contributions to the Confederate naval effort remain little known, despite the men's prominent positions. Porter especially had a hand in shaping the evolution of Confederate ironclad design.

John Luke Porter was born on September 13, 1813, in the shipbuilding center of Portsmouth, Virginia, and remained a prominent figure there all his life. His father was a master shipbuilder and shipyard owner, descended from John Porter Jr., an early settler of the Virginia Colony in the 1600s. The Porter family afterward remained distinguished for its contributions to shipbuilding and industry in coastal Virginia.[55]

From a young age Porter learned of ship construction, new technologies, and civic duty through his father, so it was no surprise when he decided to become a ship constructor. By the time of the Secession Crisis, he had served in private and government yards at Norfolk, Pittsburgh, and Pensacola; designed an ironclad (as early as 1846); and supervised construction of the US warships *Allegheny, Water Witch, Colorado* (a sister of the *Merrimack*), and *Seminole*.[56] When Virginia seceded, Porter believed that he must side with his home state even though he was against secession and favored the abolition of slavery; from the beginning, he must have been aware of the promising future he was giving up in the US Navy.[57]

Porter also made known his views on the Confederacy's unlikelihood of winning the war and lack of naval preparation: "I did not think from the beginning of the War that the Confederacy could succeed if the Federal Government chose to prosecute the War. It was a new government against an old one. We had a great many incompetent men placed in positions of trust and responsibility. We had no navy to keep our ports open."[58] Despite the fact that his beliefs clashed with the prevailing sentiments, when Virginia joined the Confederacy, Porter fully embraced his new position as a naval constructor for the war effort. He was ultimately almost single-handedly responsible for the design of virtually every type of Confederate ironclad and was promoted to chief naval constructor in 1864.[59] Thus this one man carried the development of the casemate ironclad to new heights and, along with ordnance expert John M. Brooke, shaped Navy Secretary Mallory's policies into a well-armed reality.

At age thirty-four in 1861, John Mercer Brooke was one of the ablest scientific minds in the US and Confederate naval services. He achieved some remarkable feats in ocean science by the beginning of the 1860s, although his genius for ordnance was not yet apparent. In retrospect it is remarkable that Brooke adapted to his scientific duties so well. His formal education was limited to the nine-month course given at the US Naval Academy in its early days. A tour of duty with the US Coast Survey under Samuel Phillips Lee taught Brooke much about surveying, and he learned hydrography and astronomy at the US Naval Observatory under Matthew Fontaine Maury. At the observatory Brooke invented a deep-sea sounding lead for measuring the depth of the sea at more than 2,000 fathoms, and later he won recognition for his other work. Yet nothing Brooke studied pertained to ordnance, which became his area of expertise during the Civil War.[60]

While at the Naval Observatory with Maury (who was called the Pathfinder of the Seas), Brooke had the privilege of working with one of the greatest scientific minds of his day, and in the Pacific he gained firsthand experience with the Japanese. When his ship was beached in a storm, Brooke and his crew were stranded in Japan for six months, returning to the United States on the Japanese ship *Kanrin Maru* in April 1860. It is noteworthy that Brooke not only gained the respect of his Japanese counterparts but also actually helped navigate the *Kanrin Maru* across the Pacific. Afterward he was considered a kind of expert on Japanese affairs.[61]

Brooke's loyalty to his native state of Virginia and his wife's family ultimately overrode his concerns about breaking from the Union, his lack of interest in political matters, and the fact that he owned no slaves and no Southern property.[62] Since he was well acquainted with the new Confederate navy secretary, Brooke quickly found an appointment as a naval lieutenant, the same rank he had recently held in US service.

Along with the newly commissioned Confederate Engineer in Chief Williamson and Naval Constructor Porter, Brooke was soon provided with a monumental first project: the conversion of the burned and sunken wreck of the former USS *Merrimack* into an ironclad. Besides questions of design, what to do with the existing machinery was an important consideration. In late June 1861 Brooke wrote the following in his journal:

> Yesterday Williamson and Porter, Engineer and naval constructor, arrived from Norfolk. The latter brought a model of a floating battery which he submitted to Mr. Mallory. We then discussed the matter and by unanimous consent a plan proposed by me was adopted and Mr. Porter immediately set to work on the draft. . . . I was afraid Mr. P. would, having an idea of his own, make objections to my plan, but he did not, regarding it as an improvement. Major Williamson and I then went to . . . make inquires for boilers, etc. [We] found several locomotive boilers also a new steam hammer which has never been used. . . . the boiler had however been sold.[63]

The lack of suitable boilers and components combined with the time it would take to manufacture new machinery therefore solidified the decision to repair and adapt the late *Merrimack*'s dysfunctional and waterlogged machinery. Unfortunately, Brooke and Porter's promising working relationship was later wrecked by the argument hinted at in Brooke's journal over who should receive credit for CSS *Virginia*'s design. Actually, both of these men along with Williamson contributed about equally to the great project, and all deserve credit for it.

Secretary Mallory assigned Brooke the task of developing the layout of the *Virginia*'s armor and armament in August 1861.[64] From this point on, Brooke's interest and involvement with ordnance development for the Confederate navy never abated. Brooke's experiments with armor plate and rifling old smoothbore guns led to his major triumph: the Brooke rifle. It was distinguishable by one to three wrought-iron bands shrunk over the breech to minimize the risk of bursting and to increase endurance. Brooke rifles, in 6.4-, 7.0-, and 8.0-inch calibers, and smoothbores, in 10- and 11-inch calibers, were ultimately used on virtually every Confederate ironclad by the war's end. The *Virginia*'s two pivot-mounted 7.0-inch caliber and two broadside-mounted 6.4-inch caliber were the first Brooke guns.[65]

During the war Brooke worked tirelessly on improving the design of his guns and supervising their production whenever possible. His successes led to his promotion to commander in September 1862, and he became the chief of the Office of Ordnance and Hydrography.[66] Brooke ran that bureau successfully and remained a consultant with Mallory on ironclad design until the war's end.

Although John Brooke ultimately had fairly little involvement with ironclad steam engines, his contributions to the Confederate war effort and ironclad design were impressive, and his weapon gave the Confederacy a standard naval ordnance piece that it could use for defending its harbors and arming its vessels.[67] Brooke's leadership in the vital area of naval ordnance was one of the Confederacy's greatest successes.

The man who was ultimately in charge of steam engineering for the Confederate navy remains one of its most enigmatic figures. No biography has been written on William P. Williamson, he himself wrote no memoir, and his personal records are scarce. Those facts are quite unfortunate. Unless more of Williamson's correspondence or records pertaining to his administration are located, a major gap will remain in our present understanding of the Confederate navy's steam engineering programs.

Williamson, like Porter and Brooke, was a Virginian. He was born in Norfolk on July 26, 1810, and from a young age he was involved with the new field of mechanical engineering. He studied this subject in New York as a youth and obtained employment in his hometown at the Gosport Navy Yard in 1842. Williamson was also appointed a chief engineer in the US Navy that year. He retained both his rank and his position at Norfolk (at one time working with Porter on USS *Colorado*) until the events of April 1861 caught up with him. Williamson was not prosecession or proslavery, but he valued his loyalty to his home state foremost, and when he refused to take the oath of allegiance to the United States he was imprisoned. After his release he was dismissed from the US Navy.[68]

Williamson was appointed as an engineer in the Confederate navy in June 1861, and his experience and talent ultimately made him the only real choice for engineer in chief, which he became on October 17, 1862.[69] His work on rebuilding the *Merrimack*'s machinery was invaluable to the Confederate cause in the war's early days, but like Porter and Brooke he continued to play a major part in naval affairs as the conflict progressed. After Norfolk was captured, Williamson moved to the new navy yard at Richmond, where he remained responsible for furnishing machinery plans for ships under construction as well as supervising trial and inspection runs.[70] Unfortunately, most of his bureau's papers seem to have been lost in the general destruction of the navy records at the war's end.

Williamson's familiarity with the machinery of USS *Colorado* and his skills as one of the country's foremost steam engineers made him the ideal person to overhaul and adapt the condemned, dismantled, and submerged engines and boilers of USS *Merrimack* to drive the phoenix CSS *Virginia*. It is solely because of Williamson and his team that the *Virginia*'s machinery functioned as well as it did. According to Naval Constructor Porter's correspondence, "To Engineer Williamson who had the exclusive control of the machinery, great credit is due for having improved the propeller and engines as to improve the speed of the ship three knots per hour."[71] Unfortunately, the exact modifications Williamson made are not recorded, like so many other details of his service.

Despite the overhaul of the *Merrimack*'s machinery, the inevitable breakdowns were only postponed. Although Williamson did everything he could with the materials at hand, the fact remained that the Confederacy's most famous warship was powered by defective and temperamental engines. Nobody trusted them to last very long. Second Lieutenant John Taylor Wood stated, "The motive power was the same that had always been in the ship. Both of the engines and boilers had been condemned on her return from her last cruise, and were radically defective. Of course, the fire and sinking had not improved them. We could not depend on them for more than six hours at a time."[72]

Chief Engineer Henry Ashton Ramsay reported extensively on the machinery and its gradually worsening state. Appropriately, he provided the most complete descriptions of the boiler and engine layout and his experiences in the engine room during battle.

H. Ashton Ramsay was one of the Confederacy's most gifted engineers despite his young age. He was born in the District of Columbia on July 9, 1835, but grew up in Norfolk. Ramsay served as a first engineer in the US Navy for seven years. On his return to the divided states on USS *Niagara* in May 1861, he was due for a promotion to chief engineer at the age of only twenty-five. The *Niagara* was a near sister in design to the *Merrimack*, which Ramsay had served on during that vessel's last prewar cruise. When the *Niagara* arrived in Boston the captain generously allowed any of the officers who chose to side with the seceded states to depart before the ship sailed for Hampton Roads, Virginia, and blockade duty. After a treacherous journey home Ramsay was summoned by Secretary Mallory to accept a naval commission with a view to aiding in the conversion of his former ship the *Merrimack* into an ironclad.[73]

Appointed chief engineer of the Gosport Navy Yard, Ramsay was extensively involved in the rebuilding process. His detailed testimony is quite valuable and among the most in-depth regarding any Confederate ironclad:

> I was attached to the Merrimack during her last cruise in the U.S. service when she was flagship of the squadron, therefore [I] was familiar with the machinery and fully cognizant of all its shortcomings, which I set to work to correct as far as possible, with the limited means at my command. . . . Her engines were radically defective in their condensing apparatus, the average vacum [*sic*] maintained being about one-half what it should be after the engines had worked a short time. The links, machanism [*sic*] for operating the valves and reversing the engines were also defective. The arrangement of steam chests [the main valve chamber attached to the engine cylinder] for main and expansion valves was not well designed, involving as it did a large loss of steam. In other respects the machinery was good. The main

> parts [of the] shafting, propeller, cylinders, bedplates, and c., [were] all strong and of good material; the boilers, four in number, were then known as the martin type, economical in fuel but very slow [in] combustion, and sluggish in draught. The engines were of a type known as horizontal back-acting, with two cylinders, each 72 inches in diameter and three feet stroke of piston, and as 18 pounds [per square inch] was the average pressure of steam the whole power the engine was capable of developing was only 1,000 horse. . . . The engines were all arranged for a warship, occupying very little space vertically so the highest part of them was well under the water line. With the boilers, however, the case was different. They stood very high, coming close under the berthdeck now to be the gundeck.[74]

In continuing his firsthand testimony of the *Virginia*'s crew assignments, Ramsay provided a detailed account of picking an engine room team. It remained uncertain who would be appointed chief engineer: "First, it was reported that Mr. Williamson, the engineer-in-chief, was ordered to her by the express wish of [President] Davis, who thought the engineer officer of the highest rank should have the position, but this order was revoked; two other Chief engineers were ordered to the ship, but for some reason unknown to me, the orders were also revoked. Finally, I received my acting commission as Chief Engineer (the Confederate Government was very slow in making promotions) and [was] ordered to the Merrimack as her Chief Engineer."[75]

The next step of organizing his subordinate engineers and firemen was unexpectedly arduous:

> I had some difficulty in organizing my corps of assistants, for although there were a number of Southern Engineer officiers [*sic*], who resigned and came south, most of them were older in the service than I and Ranked me, and wheras [*sic*] the complement of engineers for this class of ship consisted of two first assistants, two second assistants, and four third assistants, eight in all, I had to be satisfied with one first assistant who had been a second assistant in the old service, and two 3rd assistants taken from cival [*sic*] walks of life and who had never been to sea, five in all. . . . I selected my firemen with great care, insisting on receiving none but strong, able-bodied men, and in this was [*sic*] secured thirty.[76]

Ramsay then described fitting out the new ironclad, and he included interesting information on the trials of the machinery and his interactions with the ship's eminent commander, Franklin Buchanan. Buchanan was pressed to take his charge into action as soon as possible to score a surprise, decisive victory against the Union blockaders (all wooden ships until USS *Monitor*'s timely arrival) at Hampton Roads. This account is significant because it presents the commander of a Confederate ironclad dealing directly with matters in the engine room:

> Steam was raised and the machinery operated with the vessel tied to the wharf, a few days previous to the memorable 8th of March, when early in the morning, Captain Buchanan summoned me on deck. He questioned me closely in regard to the machinery, and as to the precautions taken to prevent the engines and boilers moving forward and the rupture of steam and other pipes in the event of a collision, possibly at full speed. After explaining in detail the methods adopted for bracing the machinery, the provision of telescopic or expansion joints to the pipes, and c., which I thought would prevent trouble to the parts referred to, he then asked if, with my knowledge of the machinery, I would be satisfied to make an attack on the fleet in Hampton Roads, without first making a trial. As yet the vessel had not been moved by her steam power since her reconstruction. . . . My reply to this was, "as it would take an hour to reach Sewell's Point Battery [held by the Confederates], we could ascertain if there was anything wrong with the machinery in this time, and I would report to him should such a contingency arise." He then remarked "very good; that he had been anxious concerning the engines as they had a bad reputation."[77]

Ramsay also mentioned Buchanan's attention to engineering details, which was unusual for a flag officer, and their discussion of signals for reversing the engines in case of a ramming event. The engine room crew was ordered to back the engines without waiting for a signal, and Ramsay was to keep Buchanan apprised of the engines' condition at all times.

Steaming down the Elizabeth River in Virginia late in the morning on March 8, 1862, for its encounter with destiny, the *Virginia* progressed satisfactorily: "The engines did very well, better than I had anticipated, and in a little while . . . the pilot told me she was going well on to 9 Knot, but I don't remember the log being heaved or any accurate measurement made of her speed."[78]

Ramsay next provided stirring firsthand testimony of the scene in his department during the battle that day and against the *Monitor* the next day. During the first engagement the *Virginia* rammed and sank the

US sloop of war *Cumberland* and burned the frigate USS *Congress*. Several other blockaders, in attempting to aid the stricken warships or flee, ran aground. Because of its deep draft the *Virginia* could not reach the stranded vessels and was forced to wait for a turn of the tide. During the night a diminutive ironclad antagonist named the *Monitor* arrived to protect the wooden ships. The stage was set for the epic first duel between ironclads the next day, recounted time and again as a pivotal episode of the Civil War. Of interest here is a steam engineer's account from the engine room during that famous battle. Ramsay wrote as follows:

> The scene in the engine and boiler rooms during the action beggared all description. Dante's Pandemonium alone approaches the weird and fiery sight here presented. The sixteen furnaces of the boilers were belching out fire and smoke, the firemen standing in front of them like so many gladiators, tugged away with devil's claw and slice bar [stoking tools], inducing by their exertions more and more intence [*sic*] heat and combustion; the noise of the crackling, roaring fires, escaping steam, and the loud and labored pulsations of the engines, made up a scene and sound that could be compared to the poet's picture of the lower regions. Then the bursting shell; several burst in the smoke stack just above the steam drum, the fragments falling on the iron fire room floor beneath. Two of the furnaces were arranged to heat shot, and out of these were being rolled those huge balls of fire, and passed up to the guns in iron cages, but notwithstanding the cannon's roar, bursting bombs, and suffocating heat and smoke, not a man quailed. . . . One shot from the enemy came near inflicting a serious injury to the machinery; it was on the second day's fight, and we supposed [it was] a solid shot or bolt from the Monitar [*sic*]. It struck us on the knuckle immediately over the outboard delivery valves of the after engine on the starboard side. The effect of the shot was to crush through the outer course of iron, and bend down the lower course into the wood backing, crush a beam down to within an inch of the stern of the outboard delivery valve of the after engine, a little further and it would have been shut up. A splinter was thrown off the beam and hurled across the engine-room with such force and velocity that it took my breath away as it passed my ear in its flight and buried itself in the iron bulkhead on the port side of the vessel. I was standing at the time with my hand on the starting valve of the after engine.[79]

The *Virginia*'s chief engineer then made further note of the battle with the *Monitor* but provided only small details about raising steam during a brief period aground, writing that "but finally, after stopping the engines for an interval, lashing down the safety valves [a very dangerous but very common practice for milking more power out of the boilers], and forcing the fires to an unusual degree, we started up again with a heavy pressure of steam and made for the Monitar."[80] Attempts to ram the *Monitor* at full power were unsuccessful because of a fear of the difficulty of extricating the ship (the *Virginia* was nearly dragged down with USS *Cumberland* the previous day).

The *Monitor* was struck at least once, but the blow was only glancing. Apparently some seams in the heavily reinforced bow were opened, but the chief engineer debunked rumors of the *Virginia*'s sinking condition after the battle from these leaks:

> I should mention, twice before he [Lieutenant Catesby R. Jones] had sent for me in regard to a report that had been made to him by the carpenter of the vessel, that she was leaking forward. I told him on each occasion that it was impossible she could be making much water as the skin of the vessel was plainly visible in the crankpits, the lowest part of the ship, and that I had the means to keep her free, even if she had a ten-inch shot in her hull, as there were two large Worthington pumps and the bilge injections of the engines which could be set to work at once in case of any such emergency.[81]

In any event, neither the *Monitor* nor the *Virginia* did quite enough damage to put the other out of action, and both retired at the end of the day to metaphorically lick their wounds. The great battle ended in a stalemate, although both sides claimed victory.

In the aftermath of these actions, Ramsay, perhaps more than anyone else, could see that the *Virginia*'s old engines had been pushed nearly to the breaking point. Little work was done on them despite his concerns. He later made a detailed report on this issue to the *Virginia*'s second and final commander, Commodore Josiah Tattnall, on April 5, 1862:

> From my past and present experience with the engines of this vessel, I am of the opinion that they cannot be relied upon. During a cruise of two years whilst I was attached to this ship in the United States service they were continually breaking down, at times when least expected, and the ship had to be sailed under canvas during the greater part of the cruise. When she [the ship] returned, the chief engineer reported that experiments to improve their working and reliability had failed, and, as the defects were radical, embracing the entire engines, recommended that they should be removed from the vessel, and such was the intention of the U.S. Government before

> she fell into our hands. The engines gave out yesterday, as I had occasion to report to you, after running only a few hours, and as I cannot insure their working any length of time consecutively I deem it my duty to make this report. At the time I was ordered to the vessel I was informed that it was not the intention to take the ship where a delay, occasioned by a derangement of the machinery, would endanger her safety, and that she would always be accessible to the navy yard for repairs; this is the reason I have deferred making this report until this time; and I also was under the impression that the Navy Department was aware of the defective nature of the machinery, and [that] her movements would be directed with a reference to this. Each time that we have gone down [into Hampton Roads] I have had to make repairs which could not have been done aboard ship very well, or, if done at all, would have required a great deal of time.[82]

Ramsay reported further on the unsuitable nature of the steam line connections and continuing repair work over the course of the next few days: "The engines of this ship are not disconnected, and one cannot be worked alone. As long as the vacuum of the forward engine holds good, the engines might be run by working the after-engine high pressure; but the vacuum of either engine is at all times precarious, and if the vacuum of the forward engine should fail, the engines would stop. Using one engine high pressure would also require a great deal of steam, which the boilers cannot generate for any length of time. The air-pump valves are now being overhauled."[83] The troubles with permanently connected engines and underpowered boilers were a problem that ultimately plagued several other Confederate ironclads.

Chief Engineer Ramsay later moved on to another position in which he made significant contributions to the war effort, especially regarding naval steam machinery. The Charlotte Navy Yard in North Carolina and Ramsay's skillful management of that facility are discussed in chapter 5.

Although testimony regarding the *Virginia*'s machinery compared to other ironclads is certainly not lacking, one other engineer's valuable words in active service should be presented here. Third Assistant Engineer Eugenius Alexander Jack recorded several interesting details of serving in the engine and fire rooms during his ship's battles. Jack's account is especially interesting because he was one of the lower-ranking officers, whose accounts are comparatively rarer than those of captains and even chief engineers.

Fortunately, Jack's memoir has been published at the behest of his granddaughters. Jack was about eighteen when the Civil War began; he had finished schooling just two years earlier. According to his memoir,

Jack was an apprentice at Gosport Iron Works and gained some firsthand engine room experience on the Bay Line steamer *Adelaide*. He was at Norfolk when Virginia seceded, and he chose loyalty to his home state. After serving for a brief time in the Virginia army, Jack was detailed to the Gosport Navy Yard, where his skills as a machinist and his engineering experience allowed him to participate in the work on the *Merrimack*. The young man's interest in steam engineering was rekindled:

> I now thought it was time to make my effort to get an appointment in the Confederate navy, and duty on her [the *Merrimack*] if possible. It was probable though that if I got my warrant as engineer I would be ordered to the "Merrimac," for I had been working on her engines and the desire of the Government would naturally be to put those in the engine room of whom most might be expected. . . . I made my application through Chief Engineer Williamson and in due time was fortunate enough to receive my warrant as Third Assistant Engineer in the C.S. Navy, and to be ordered to the "Merrimac" now called the "Virginia," but except in official communications, ever called by the former name.[84]

Jack described in detail this most unusual first billet and provided a list of the engineers on board and their duty stations during the *Virginia*'s two battles. This is valuable because it is unusual to have the names of all the engineering officers for any Confederate ship. One of these men, John Tynan, will be encountered later in this work in connection with CSS *Savannah*. The list of the *Virginia*'s engineers was given by Jack as follows:

> Chief Engineer H. A. Ramsey [*sic*]
> 1st. Asst. Engineer John Tynan in engine room
> 1st. Asst. Engineer Loudon Campbell in engine room
> 2nd. Assistant Engineer Benjamin Herring (I think in fireroom and in charge of auxiliary steam pumps)
> 3rd. Assistant Engineer E. A. Jack in fireroom
> 3rd. Assistant Engineer E. V. White at the bell to signal to the engine room[85]

Jack's impressions of the machinery itself also corroborate Ramsay's:

> Her motor power was two horizontal condensing engines, which were supplied with steam from four Marlin [Martin] boilers containing three furnaces each. The engines operated by link-gear, which was shifted by the arc and pinion movement and were quite difficult to

> manage, and required at least two men at each reversing gear. For one of the furnaces, a frame of wrought iron was made upon which shot for smooth bore guns could be placed and heated and tool and trivets were supplied to handle and carry them to the guns. There were two auxiliary steam pumps of the Worthington type, with about six-inch cylinders. These pumps could be used either to supply the boilers with water or to pump water from the bilge.[86]

Jack also provided some useful details of what it was like in the machinery spaces during the fight of March 8, 1862, against the Union vessels at Hampton Roads: "The preparation for the fight now began on our ship and all hands were called to quarters, and I sought my station down in the fire-room twenty feet under water mark. . . . The suspense was awful, but it was relieved occasionally, by some information through the ash chute that my comrade [Third Assistant Engineer E. V.] White would communicate."[87] In this manner, Jack and the other men in the fire room were apprised of the situation outside and learned of the *Virginia*'s incredible victory over the blockading fleet that day.

The *Virginia*'s monumental battle with USS *Monitor* on March 9 was recorded in some detail by Jack, but he had relatively little to say about operations in the machinery spaces. Like Ramsay, he briefly related the shell hit near the waterline: "One shot struck directly over the outboard delivery, and because that was a weak spot [it] broke the backing of the shield and sent a splinter into the engine room with about enough force to carry it halfway across the ship."[88]

Again like Ramsay, Jack made note of rumors after the battle that the *Virginia* was forced to retire because of bad leaks from ramming the *Cumberland* the previous day and attempting to do the same to the *Monitor*: "As I was in the fire-room within sight of the steam bilge pumps at all times, in fact sitting by them often, I know that they were not required to do any extraordinary duty and that no alarm about the sinking or leaky condition of the ship was felt in that section."[89] After this tidbit Jack gave no more technical details of the *Virginia*'s engine spaces.

Eugenius A. Jack continued to serve on the *Virginia* during the ship's uneventful final weeks. When the ironclad was destroyed, the crew marched to Suffolk, Virginia, then to Richmond, where the engineer was ordered to report to an ironclad construction site in Memphis, Tennessee.[90] He could not know that this vessel too was destined for the forefront of Civil War naval history. Jack's valuable testimony will therefore be presented again in the next chapter.

Neither the *Virginia* nor the *Monitor* ever fought again. Despite the Confederate ship's efforts to press an engagement, the *Monitor* remained at its post guarding the wooden blockaders. In its final days the *Virginia* was a much improved ship, with a new ram, iron gunport shutters finally fitted, and additional armor added below the waterline, but it still retained too deep a draft to allow its escape up the James River when the Union captured Norfolk in May 1862. Commodore Tattnall decided to destroy his own ship rather than risk it falling into enemy hands, thus beginning a trend that affected most Confederate ironclads at the end of their careers.

There is actually a large piece of the *Merrimack*'s, or *Virginia*'s, machinery preserved today. One section of the propeller shaft was recovered sometime after the ship's final destruction. It now rests outside the entrance to the Museum of the Confederacy in Richmond, Virginia, and is an impressive sight. A reminder of this artifact's historical importance is a simple commemorative plaque of the Battle of Hampton Roads stamped on top.

CSS *Baltic*

One of the most obscure Confederate ironclads is the converted sidewheel towboat CSS *Baltic* of Mobile, Alabama. Little information exists about this craft, which was the only armored vessel to serve on Mobile Bay for more than a year. Fortunately, some details on the machinery do exist, mostly because of a Union survey of the captured vessel at the war's end. Not many records concerning the *Baltic* have been found, and Confederate documents relating to the vessel are almost nonexistent.[91]

A large portion of the extant information on the *Baltic* comes from the Gunboat *Baltic* Papers compiled in 1938 by Clyde E. Wilson and held in the Alabama Department of Archives and History. In addition to the Union survey report, relevant technical details about the steam machinery can be found in newspapers dating from the vessel's launch.

The *Baltic* was originally built for Captain William Bragdon in 1856 in New Albany, Indiana.[92] The shipyard of William Jones constructed the hull, and the large firm of Lent, South & Shipman furnished the machinery. This foundry was described as having a "large blacksmith shop [which] occupied the entire river front from Pearl to State street and from the Water street front to [the] Southern Railway track. This firm made all [its] own coke. [Its] principal business, however, was the building of steamboat machinery of every description, and much of the time [the firm was] compelled to run day and night."[93] Specifically regarding the *Baltic*, a contemporary article noted that "all . . . appurtenances are constructed in that substantial and workmanlike manner which distinguishes the work of this firm."[94] The power plant itself was typical for a paddlewheel-driven ship. Unfortunately, it was not capable of moving the *Baltic* after conversion at more than a man's walking pace.[95]

The *Baltic* was described in multiple newspaper accounts as a side-

Figure 2.6. Contemporary engraving of the *Baltic*. The level of accuracy is uncertain because there are few existing descriptions of the vessel as converted. The configuration of the smokestack and cowl vents, at least, agrees with the boat's previous incarnation as a riverboat (boilers far forward).

wheeler 176.5 feet in length, 37.3 feet in beam, and 7.6 feet in depth of hold, displacing about 450 tons.[96] It was taken to Mobile, Alabama, by Bragdon upon completion and served its intended function as a towboat and cotton lighter on Mobile Bay until its 1861 purchase by the state of Alabama.[97]

After Alabama's secession, a state committee was formed for the purchase of a vessel to be converted into an ironclad. By December 13, 1861, the committee had followed the required specifications for a suitable vessel and purchased "the good Steamboat called 'The Baltic,' her boilers, Engines Machinery, tackle, Apparel and furniture" for the sum of forty thousand dollars. Conversion of the small lighter into an ironclad commenced on December 21, 1861, and was finished the next May.[98] Little is known of the exact layout other than that it was cottonclad aft (fig. 2.6).[99]

The *Baltic*'s dimensions after conversion were increased to 186 feet in overall length, 38 feet in beam, 7 feet in draft, and a displacement of 624 tons. The boat's bow was lengthened and strengthened into a ram, and four heavy guns were equipped. All this extra weight, combined with that of the armor, left the *Baltic* very slow and unmanageable. To compound the problem, rottenness set in, and the ironclad was relegated to sideline duties only.[100] The boilers were also described as being out of repair. The few Confederate accounts clearly reveal that service on the ironclad was uncomfortable at best, and no one had much faith in its abilities.

The *Baltic*'s wartime career was uneventful. Its first captain was Lieutenant James D. Johnston, who was later given command of the newly completed CSS *Tennessee*.[101] He sailed that ship to glory under Franklin Buchanan at the Battle of Mobile Bay (August 5, 1864). After Johnston's transfer (a major promotion, in many ways), Lieutenant Charles Carroll Simms was given command of the *Baltic*. Apparently by that time, late 1863 and early 1864, the boat's general condition had deteriorated so much that Naval Constructor Porter was called in to make a structural examination.

On March 20, 1864, after Porter had surveyed the *Baltic*, Simms wrote, "Porter, the naval constructor, has made a very unfavorable report of the condition of the ship and recommended that the iron be taken from her. . . . the *Baltic* is as rotten as punk, and is about as fit to go into action as a mud scow."[102] Therefore the *Baltic* did not take part in the Battle of Mobile Bay. This must have been extremely frustrating to a man like Simms, who by 1863 had accumulated a formidable record of varied experience, becoming one of the Confederacy's foremost junior naval officers.

Charles Carroll Simms was born in Alexandria, Virginia, in 1811. At the age of twenty-eight he joined the US Navy and served until his home state seceded. His services were greatly needed for the fledgling Confederate forces; he was commissioned in the state navy of Virginia as a lieutenant and became a first lieutenant in the Confederate navy in June 1861. Simms served with distinction on CSS *Virginia* as the bow gun commander during the Battle of Hampton Roads, but he moved to CSS *Nansemond* of the James River Squadron after the *Virginia* was destroyed. Before his appointment to the *Baltic*, Simms was involved with the commerce raider CSS *Florida* during that ship's visit to Mobile, and later he became the commander of the giant ironclad *Nashville*. He served briefly at the Selma Naval Cannon Factory in the war's final months, and he ultimately surrendered with the remainder of the Confederate naval forces in the Mobile area in May 1865.[103]

Although it is easy to imagine Simms's opinion of the *Baltic* in general, he left no account of his decrepit command's machinery. Neither did any of the ironclad's engineers. However, abundant engineering data was provided by other sources as the *Baltic*'s condition worsened along with the Confederacy's fortunes in the last year of the war.

In the final months, Flag Officer Randolph Tucker wrote a report on the unsafe condition of the boilers: "The boilers of the steam ram *Baltic* were pronounced by the engineers as unsafe, and were being patched and strengthened."[104] This was after the deteriorating vessel's armor had been removed (both for safety purposes and for supplying the new ironclad CSS *Nashville* with much-needed iron plate). Not long afterward the *Baltic* was decommissioned. At the war's end, it was sent up the Tombigbee River in Alabama with the *Nashville* and several other Confederate vessels and subsequently surrendered to federal authorities at Nanna Hubba Bluff on May 10, 1865. A Union survey of the ship in June reported the following:

> The *Baltic* is built like all other river steamers, but was strengthened when she was constructed into a ram. Her hull is in good condition below the load line, but above the load line her hull and upper works are rather rotten in some places. Her main deck is in a very bad condition and rather rotten. Her cylinder timbers are affected with the dry rot in some places, and not very well secured. . . . These engines are in good condition and repair, with the exception of the timberwork. . . . This ship is not safe to run in consequence of the bad condition of the boilers.[105]

The boiler assembly was next reported on in detail. The *Baltic* was powered by four "very old and leaky" western river–style horizontal double-flue boilers. Each was 24 feet long and 3 feet in diameter (or 28 feet long and 40 inches in diameter, according to the 1856 newspaper articles); the flues were 15 inches in diameter.[106] The furnaces could burn either wood or coal, in keeping with standard practice of the day, and the boat could carry 75 tons of fuel. The *Baltic* also carried a doctor engine for pumping and boiler feedwater purposes, which was also standard.[107]

The main propulsion machinery was in good condition. The *Baltic*'s two engines were described as inclined noncondensing, "of the usual Western river type and arrangement." The cylinder diameter was 22 inches with a 7-foot stroke, each independently turning a 29-foot-in-diameter, 8-foot-wide paddlewheel.[108]

It was initially recommended by the Union surveyors that the *Baltic* would once again make an excellent towboat with repairs to the hull, woodwork, and boilers. Nothing ever came of the proposal, however, and the worn-out vessel was eventually sold.[109]

Conclusion

The Confederacy's first three ironclads differed greatly in design and service life. A common thread to all, however, was their poorly performing machinery. Although the *Manassas*'s and the *Baltic*'s engines and boilers were suitably powerful and performed well during those vessels' previous careers, they were not adapted to efficiently drive heavily timbered and armored warships. The *Manassas*'s use as a ram and the *Baltic*'s lack of maintenance certainly did not help, either.

The *Virginia*'s steam plant was a different story. Although the boilers proved mostly satisfactory, the engines were known to be of poor design from the time they were new. Not much could be done with them short of total replacement, even though a large number of the Confederacy's leading engineers, constructors, scientists, and naval officers were involved at one time or another with the ship's conversion and fitting out. Engineer in Chief Williamson's modifications to the engines probably only delayed their inevitable total failure, but the *Virginia*'s machinery performed amazingly well during its historic battles.

Even though some of the nation's leading industrial establishments built the machinery used by the ironclad conversions, manufacturer reputation and skill could by no means guarantee adequacy. This was symptomatic of the early days of steam engineering generally, coupled with the fact that the *Enoch Train*, the *Baltic*, and the *Merrimack* were adapted for a purpose their designers could not have anticipated. The skill of the Confederate engineers who modified and operated them certainly contributed in great measure to any success they did have as ironclads.

The conversion of these vessels did answer the Confederacy's dire need for armored steam warships at the beginning of the Civil War. The mobilization of Southern industry seemed to promise better purpose-built hulls and engines for the next generation of ironclads.

THREE

EARLY NONSTANDARD DESIGNS

All depends upon the principle of construction which remains to be tested.
—Lieutenant Robert D. Minor, Confederate States Navy

The Confederacy's next group of ironclads benefited from keel-up design and construction but also suffered from a muddled idea about purpose. At the beginning of the Civil War an ironclad's role, whether in harbor defense or open-water offensive operations, was not fully defined. Therefore the first ironclads conceived as such attempted to combine attributes of both riparian and coastal warships. Because of this uncertain design philosophy, compromises had to be made, and the *Louisiana*, the *Arkansas*, and the *Georgia* all suffered from serious weaknesses. Machinery troubles were once again at the forefront. (CSS *Mississippi* also belongs to this group but was not completed. That singular vessel will be discussed in a later chapter.)

CSS *Louisiana*

What became one of the Civil War's largest ironclads was begun on October 15, 1861, just north of New Orleans. This vessel was one of the strongest and most powerful ironclads built by the Confederacy, but it was rendered a failure by its unique propulsion system. CSS *Louisiana* had several features that could have given good service, but it was ultimately hampered by a lack of suitable machinery, experienced workmen, and materials.

The *Louisiana* was designed and built by a Kentuckian, E. C. Murray, an early proponent of ironclads and a veteran of twenty years' experience building sail and steam vessels. He was one of the first to submit plans for an ironclad vessel in April 1861, and later he claimed to have designed the *Virginia*, saying that Lieutenant Brooke had stolen his plans.[1] No evidence for Murray's assertion has ever been found. By late 1861, however, he had created a new design and was poised to begin construction on an ironclad intended for the defense of the vital port of New Orleans.

New Orleans probably contained the greatest manufacturing capability of any city in the Confederacy other than Richmond before its capture in April 1862. The manufacture of steam machinery, particularly for riverboats, was a thriving business by the 1860s because the city was uniquely situated on the Mississippi River, functioning as both a river port and an ocean port. New Orleans had many private businesses devoted to the manufacture of steam engines. Foremost were the foundries of Leeds & Company; Clark; Bennett & Lurgis; Shakespeare; Gretna; John Armstrong; and Kirk & Company.[2] Abundant resources for shipbuilding existed in the area in addition to these establishments, and across the river in Algiers were eight drydocks. It appeared that ironclads constructed in the Crescent City (as New Orleans was known) could be built well and quickly.

The *Louisiana* was the first ironclad actually laid down and constructed for the New Orleans station, on a new shipyard site at Jefferson City, Louisiana. Murray's conception called for an extremely simple and bargelike shallow-draft hull with minimal curves, a flat bottom, and straight sides. Taking its design cues from the *Virginia*, the hull was unarmored and mostly submerged for protection; only large wooden breakwaters and a massive 45-degree-angle casemate showed above the surface. The casemate was formidably pierced, originally, for twenty-two guns, although only sixteen were ultimately carried.[3]

Despite Murray's extensive experience with steamboats and their simple machinery, his ideas for the *Louisiana*'s power plant were even more unusual than the vessel's construction. The propulsion setup of two centerline paddlewheels and two propellers was not just strange but also extremely troublesome. Ostensibly the primary reasoning was protection from battle damage for the wheels, with the small screws to aid in maneuvering.[4] The *Louisiana*'s lack of motive power caused by problems with this unique layout plagued it throughout its career and eventually contributed to its loss.

Most of the *Louisiana*'s short service life proved to be an odyssey of assembling and fitting the disparate propulsion systems to the vessel. During construction it was decided to use machinery from other boats instead of manufacturing it new, primarily to save time. The industrial facilities of New Orleans were already overwhelmed in turning out weapons and other wartime commodities. Two riverboat and two propeller engines therefore needed to be procured for Murray's ironclad.

Half of the *Louisiana*'s machinery came from the large river steamer *Ingomar* of the Memphis & New Orleans Packet Company.[5] The boat was a sidewheeler built at Louisville, Kentucky, in 1854. Launched on August 8 of that year at the shipyard of Samuel King, the *Ingomar* displaced 730 tons, was 275 feet in length and 40 feet in beam, and had a draft of 7.5

feet. Slightly different dimensions were given in a newspaper description: 280 feet long, 37 feet in beam, and 9 feet in depth of hold.[6] The *Ingomar*'s two inclined high-pressure engines had either a diameter of 28 inches with a 7-foot stroke (according to Frederick Way) or a diameter of 30.5 inches with a 9-foot stroke (according to the *Louisville Daily Courier*). The engines were built by Roach & Long of Louisville. The boilers were listed as five in number, 33 feet long and 42 inches in diameter.[7]

It is fairly unusual that the firm of Roach & Long is shown in the historical record as having produced the *Ingomar*'s machinery, and it is even more unusual that something is known about it. The construction histories of most steamboat power plants are largely forgotten. Many steam engine manufacturers (especially smaller establishments) remain unknown. The Civil War came during the heyday of the steamboat era, when thousands of boats plied the western waters and hundreds of local firms supplied their machinery. Often a boat was built in one place, then taken to another for the fitting out and installation of machinery, making specific firms difficult to trace.[8] The history of Roach & Long was probably fairly typical of the time. The establishment was founded in the 1840s and built many steamboat and stationary engines before turning exclusively to the manufacture of water and gas main pipes after 1860.[9]

On the *Louisiana* each *Ingomar* engine ostensibly turned a paddlewheel that was 27 feet in diameter and 19 feet wide, although these dimensions may have been reduced to a smaller 20 feet in diameter for the fore wheel and 24 feet for the aft wheel.[10] The main paddle shafts were each 24 feet long and 13 inches in diameter.[11]

Most contemporary accounts of the *Louisiana* come from the 1863 *Report of Evidence Taken before a Joint Special Committee of Both Houses of the Confederate Congress to Investigate the Affairs of the Navy Department*, which investigated the spectacular loss of New Orleans in April 1862 along with two of the Confederacy's largest ironclads (CSS *Mississippi* in addition to CSS *Louisiana*). Parts of these testimonies describe the efforts to install the *Louisiana*'s machinery. E. C. Murray took the witness stand on February 20, 1863, and stated the following: "I bought the steamer *Ingomar*, and transferred her machinery to the *Louisiana*. I . . . contracted with Kirk [& Company of New Orleans] to build me two propellers and two propeller engines. Not being able to get them anywhere else, I applied to Leeds [& Company of New Orleans] to construct them, but he would not touch them. They [Kirk] were engaged in removing the machinery from the *Ingomar* to the *Louisiana* about two months, and about the same time building her propeller machinery."[12]

The small engines and shafts required for the propellers may have been purpose built by Kirk & Company, but better evidence suggests they were removed from tugboats and adapted to the *Louisiana*'s specifications. In any case, the larger establishment of Leeds & Company was too loaded down with wartime contracts to build any of the *Louisiana*'s machinery. Kirk's smaller establishment ultimately agreed to furnish propeller engines and shafts to Murray. The history of this firm unfortunately remains obscure. Such is the case with most of those in New Orleans during the war.

No information specifically shows that the propeller engines were built new by Kirk & Company. It instead appears more likely that engines were taken from tugboats with machinery built by Reanie, Neafie & Company, one of the greatest engine works in the United States. That firm also manufactured steam plants for several other prewar vessels that ended up powering Confederate ironclads.

That giant establishment, commonly known as Penn Works, was started in 1838 by Jacob G. Neafie on the corner of Germantown Road and Second Street in Philadelphia, Pennsylvania. Gradually, Neafie's small collection of wooden sheds and small shops grew into one of the largest shipbuilding and machinery manufacturing centers in the country. Extensive works covering nearly ten acres contained nearly every type of equipment necessary for building engines, boilers, and iron vessels.

Engines were completed for many merchant vessels and warships of all types, including the frigate USS *Lancaster* and several gunboats. During the Civil War, "the engines for about one hundred and twenty vessels, of all classes, were built here, some of them among the largest in the service."[13] A large part of this success was a result of the company's extensive experience in forging propellers.[14] By 1856 Reanie & Neafie had built more propeller engines than any other manufacturer in the United States, and it even had the patent rights for the curved propeller, which was the originator of most modern propeller designs.[15]

The single-cylinder inverted direct-acting variety of propeller engine was very common by the 1860s, and Reanie, Neafie & Company built the majority of them. Many were installed on tugboats. At the beginning of the Civil War, any tugboats in Southern waters became a valuable resource both as working vessels and as sources of marine machinery for the Confederate navy.

A witness to the *Louisiana*'s trials reported that "two engines from little tugs were placed aboard, one in each of her sterns."[16] This statement is the only direct bit of information on the source of the *Louisiana*'s secondary propulsion. Indirect evidence points more specifically to the tugboats *Decatur* and *America*: both were constructed in the Philadelphia area, and both received their engines from Reanie, Neafie & Company.[17] These are the only screw-propelled tugboats known to have existed in antebellum

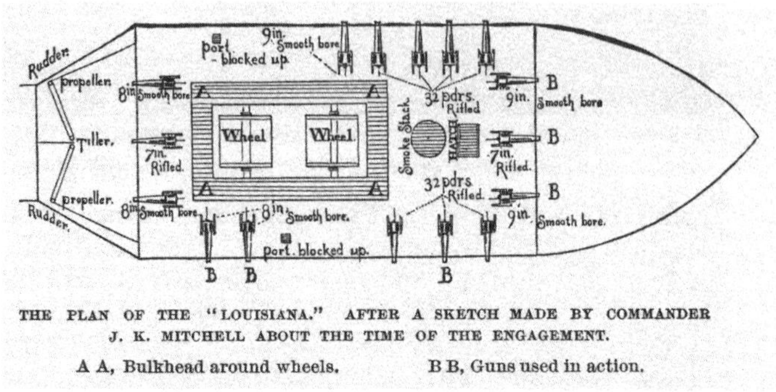

Figure 3.1. Rough plan of the *Louisiana* based on a sketch by Commander John K. Mitchell.

New Orleans that are not accounted for either by military service or by postwar registration. In addition, more than half of the ownership of each was by Northern parties, which made them likely to be seized.[18]

The known facts of the *Decatur* and the *America*'s construction and machinery also follow these conclusions. Both vessels had nearly identical low-pressure vertical-inverted single-cylinder engines with an 18-inch diameter and stroke.[19] The *Decatur* was launched in 1855 and measured 63.4 feet long, 14.9 feet in beam, and 5.3 feet in depth, with a 45-ton displacement. The boat and its machinery were completed by Reanie, Neafie & Company. The *America*'s hull was constructed by Jacob Birely of Kensington, Pennsylvania, in 1857, but its machinery was also furnished by Reanie & Neafie. The dimensions given for the *America* were 64.9 feet long, 17.2 feet in beam, and 7.1 feet in depth, with a 70-ton displacement. Both of these very similar boats were serving in New Orleans at the outbreak of the war.[20]

The above information is convincing, but nothing is definitely known except that Kirk & Company was responsible for furnishing the engines to turn the *Louisiana*'s small five-feet-in-diameter screws via wrought-iron shafts 6.5 inches in diameter and 18 or 19 feet long.[21] They were not completely installed until immediately before the ship's destruction. Providing steam to this machinery (as well as to the paddlewheel engines) were the five horizontal high-pressure boilers taken from the *Ingomar*. That vessel's doctor engine and boiler were also installed in the *Louisiana*.[22]

It is noteworthy that the *Louisiana*'s completed wrought-iron shafts did not fit properly. Murray had to persuade Kirk to construct a tilt hammer, "by which means he succeeded in drawing them out and adapting them to my purpose. The only difficulty I had was to get him to get along with the work."[23] Despite the firm's slowness in turning out materials, it seems they had the machinery and capability to do the job.

Despite various troubles, the work slowly progressed and the *Louisiana* was finally ready for launch on February 6, 1862. The huge ironclad's dimensions were 264 feet long, 62 feet in beam, and 6 feet in draft, with a displacement of 2,751 tons.[24] No plans of the *Louisiana* or its machinery layout are known to exist, but a rough sketch of the ship's general configuration was made by Commander John K. Mitchell, who replaced Captain George Hollins in commanding the naval forces at New Orleans. The drawing depicted in Figure 3.1 forms the basis for most conclusions on the *Louisiana*'s unusual propulsion systems and main deck layout.

John Kirkwood Mitchell was another long-serving US officer who had gone South. He led the Confederate naval forces at the Battle of Forts Jackson and St. Philip below New Orleans (in which CSS *Manassas* figured so prominently) but was captured in the Confederate defeat. After being exchanged, he went on to command the Office of Orders and Detail in Richmond before returning to ironclads as the commander of the James River Squadron from 1864 until the war's end. After the war Mitchell lived in Richmond until his death at age seventy-eight in 1889.[25]

Commander Mitchell's later testimony during the investigation of the Confederate Navy Department's conduct before the loss of New Orleans provided valuable firsthand information on the arduous process of completing the *Louisiana*'s machinery. There was, in reality, little he could do to hurry things along, even as events moved toward a Union attempt on the Crescent City. Other accounts corroborated Mitchell's assertion that everything had been done that could be done.

A trial run on April 19, 1862, shortly after the *Louisiana*'s launch revealed for the first time the weakness of the ironclad's propulsion system: "It was found that it was with difficulty she kept up with the current when going downstream, and when pointed upstream she was carried down at the rate of two or three knots an hour. [After the ship was] Towed back to the wharf, two engines from little tugboats were placed aboard, one in each of her sterns. This increased power was not perceptible."[26]

The *Louisiana* was still incomplete because of delays in construction and machinery assembly when it was ordered 60 miles downriver to Fort St. Philip to prepare for defense against Flag Officer David Farragut's approaching fleet. Commander Mitchell later testified, "To have prepared her fully for a fight, it would have taken at least six weeks, if not two months longer. In work of that kind there are many causes of delay. For instance, we were a long time waiting for the propellers. The engineers were promising every day that they would be ready, but they failed to come up to their promise, though I am satisfied it was made in good faith. The work was much more arduous and difficult than they apprehended."[27] The *Louisiana*'s motive power was only partly assembled when the deployment order came. The giant ironclad was incapable of stemming the Mississippi current with just paddlewheels. According to the ship's third lieutenant, William C. Whittle Jr., "She had two propellers aft, which we never had an opportunity of testing. . . . The vessel left New Orleans on the 20th of April, I think. The work on the propellers was incomplete, the machinists and mechanics being still on board. . . . The center wheels were started, but were entirely inefficient."[28] The *Louisiana* had to be towed down to a mooring adjacent to Fort St. Philip. Several tugboats were required to assist the unwieldy 2,700-ton ironclad.

Workers continued to feverishly assemble the *Louisiana*'s propeller engines and shafts, but the Union attack came before their completion. On April 24, 1862, Farragut's fleet stormed past the forts and proceeded upriver to take New Orleans. Most of the Confederate fleet was destroyed, but the *Louisiana* resisted the heaviest federal guns without damage and remained tied to the riverbank. Because of the defects in the gunports' size and layout, it is uncertain how much the vessel's few trainable guns contributed to the melee. The ironclad's captain, Charles Flemming McIntosh, was the only noted casualty on board. While McIntosh was observing the battle from atop the shield, his legs were taken off by a shell, and he later bled to death.[29]

McIntosh's appointment as the *Louisiana*'s master was an interesting twist of fate and warrants a brief digression. McIntosh was a Norfolk resident and an experienced navy man when he resigned from the US service in 1861. He was appointed a commander in the Virginia navy and served at the Gosport Navy Yard when USS *Merrimack* was being converted into CSS *Virginia*.[30] While there, he was able to observe the conversion process and talk with engineer H. Ashton Ramsay. Ramsay reported the following:

> My dear old friend and comrade Captain Chas. MacIntosh [*sic*], on waiting orders, used to come over & stand on the granite curbing of the [dry] dock, and watch the work as it made its snail-like progress. He took great interest in the ship, but had very little faith in her success. I shall never forget his last words when he left me on the eve of starting to take command of the ram [the *Louisiana*] at New Orleans: "Good-bye, Ramsay, I shall never see you again; she [the *Virginia*] will be your coffin." Hardly two weeks [months] after this, the poor fellow had both his legs shot from under him . . . killing him.[31]

Thus a man who sincerely doubted whether an ironclad would even float finished his career and life as the captain of one (even more experimental than the *Virginia*, in many respects). If McIntosh had remained inside the *Louisiana*'s casemate, he would probably not have become the only Confederate ironclad captain killed in combat.

Even as Farragut's fleet captured New Orleans, mechanics and engineers on the bypassed *Louisiana* raced to complete its machinery in the hope of escape. By April 28, 1862, the work was finally done. It was found that the *Louisiana* would not move even with full steam pressure and both wheels and propellers operating.[32] The huge ironclad was subsequently destroyed to prevent capture when the nearby forts finally surrendered to shore-based Union forces.

Chief Engineer Wilson Youngblood, reporting to Commander Mitchell even as a prisoner in August 1862, had a final word on the failure of one of the Confederacy's strongest ironclads:

> SIR: I respectfully report the machinery of C.S.S. *Louisiana* all in good working order on the morning of April 28, 1862, just before it became necessary to destroy the vessel. But I do not think it would have been able to handle the vessel, the wheels being put in the middle of the vessel, one right abaft the other, so that the after wheel could do no good whatever. And again, when the wheels were working, they would force the water out under the stern so that it would form an eddy around the rudder so that she would not steer, and if we tried [to] steer her with the propellers, she could not stem the current. Consequently she was unmanageable in the Mississippi River.[33]

Archaeological investigation of the *Louisiana*'s wreck site may someday uncover more information on the finished configuration and machinery layout of this unique warship.

CSS *Arkansas*

Only one other Confederate ironclad approaches CSS *Virginia* in level of fame. CSS *Arkansas* was intended to defend the Confederacy on the upper Mississippi River. During its brief service life, the *Arkansas* fought more battles against overwhelming odds than any other Civil War ironclad. Fortunately, abundant information on the construction, machinery installation, and exploits of this remarkable vessel is preserved in the accounts of its equally remarkable master and officers.

The *Arkansas* and its uncompleted sister ship the *Tennessee* (not to be confused with the later, more famous *Tennessee* of Mobile Bay) were intended from the outset as offensive river and seagoing warships.[34] As such, the *Arkansas* combined a heavy armament with several hybrid design features. These advantages eventually all came to naught when the *Arkansas*'s unreliable engines directly caused its loss.

Construction on the two ships began in October 1861 at a landing near Fort Pickering below the strategically important Mississippi River town of Memphis, Tennessee, under contract to steamboat builder John T. Shirley. On April 24, 1862, the first of the two, christened the *Arkansas*, was launched and hurriedly put in preparation for towing to a safer location. The fall of New Orleans that very day and the approach of a federal gunboat fleet upriver made imperative the removal of the *Arkansas*'s incomplete hull to a hidden location in the swamp. Just two days later, the sister ship *Tennessee* was burned on the stocks as Union naval forces attacked Memphis.[35] Through this turn of events the *Arkansas* was saved and towed far up the Yazoo River to Greenwood and Yazoo City, Mississippi.

At the new site, construction lagged for want of materials and skilled workers and because of pusillanimous leadership. All that changed when Lieutenant Isaac Newton Brown was ordered to take over command of the vessel. He soon proved himself more than equal to the challenges of superintending construction before leading the *Arkansas* into battle.

Brown was a highly regarded officer with twenty-seven years of naval experience when he was ordered to Yazoo City. Brown, a Mississippian, was commissioned a lieutenant in the Confederate navy on June 6, 1861, but during the first year of the war he made significant contributions in aiding Confederate army forces in the west. He also supervised the construction of several gunboats on the Tennessee River that were never finished (much of the area was taken by federal forces early on). It appears that Brown was unusually skilled in locating, collecting, and organizing available resources, a valuable talent in the always materiel-starved navy. For these reasons Brown was given the formidable task of completing the *Arkansas*. He was well rewarded for fulfilling this duty when he was made commander.[36]

Brown arrived at Yazoo City on May 29, 1862, and promptly got things moving despite all the difficulties of working in such a remote location.[37] With his herculean effort of willpower and organizational skill, rapid progress was made, although the initial picture was chaotic. Among other problems, a major issue was that the *Arkansas*'s engines were in place but not connected. In fact, the whole ship was a cluttered mess. Brown wrote, "Her condition was not encouraging. The vessel was a mere hull, without armor; the engines were apart; guns without carriages were lying about the deck; a portion of the railroad iron intended as armor was at the bottom of the river, and the other and far greater part was to be sought for in the interior of the country."[38]

Somehow overcoming all the obstacles and obtaining both iron armor and guns, Brown succeeded in speedily completing the *Arkansas*. One wonders if anyone else could have done it. Though justifiably proud, the commander also recognized his vessel's deficiencies, especially the engines. In the end these proved to be the ship's greatest weakness:

> Our engines' twin screws, one under each quarter, worked up to eight miles an hour in still water, which promised about half that speed when turned against the current of the main river. We had at first some trust in these, not having discovered the way they soon showed of stopping on the center at wrong times and places; and as they never both stopped of themselves at the same time, the effect was, when one did so, to turn the vessel round, despite the rudder. Once, in the presence of the enemy, we made a circle, while trying to make the automatic stopper [a device designed to make both engines stop at the same time in case one broke down; apparently it never worked, for it is never mentioned again] keep time with its sister-screw.[39]

The *Arkansas*'s machinery is forever linked with the failure of that otherwise well-founded vessel. It is surprising that the ironclad held together as well as it did during its many battles (considering what Brown had to work with), but it is even more surprising that the *Arkansas*'s new and purpose-built engines and boilers should have functioned so poorly.

The machinery for the *Arkansas* was built in Memphis by a local firm on Adams Street. Unfortunately there is some confusion about the identity of the company. An 1860 Memphis city directory lists the Union Foundry & Machine Shop, but at least one other source lists the engines as coming from the foundry of S. M. Coates.[40] Almost nothing is known about these facilities; they may even just be different names for the same firm. More research must be done to clear up this mystery.

In addition to the engines and boilers produced in Memphis, the *Arkansas*'s propeller shafts were forged and turned by Leeds & Company in New Orleans. Leeds far exceeded any other New Orleans company in output and experience. The firm was founded in the 1820s and was the world's largest maker of steel cotton-bale presses by the time of the Civil War. Leeds also had experience in casting, forging, and finishing iron components for sugar and cotton equipment and steam engines, employing a force of more than four hundred workers. During the war, production was shifted to manufacturing arms for the Confederate armies.[41] The *Arkansas*'s shafting was ultimately Leeds's only real contribution to Confederate ironclad construction. The firm quickly became too busy to aid in the vital work being carried out on the two large ironclads (the *Louisiana* and the *Mississippi*) intended for the defense of New Orleans itself.

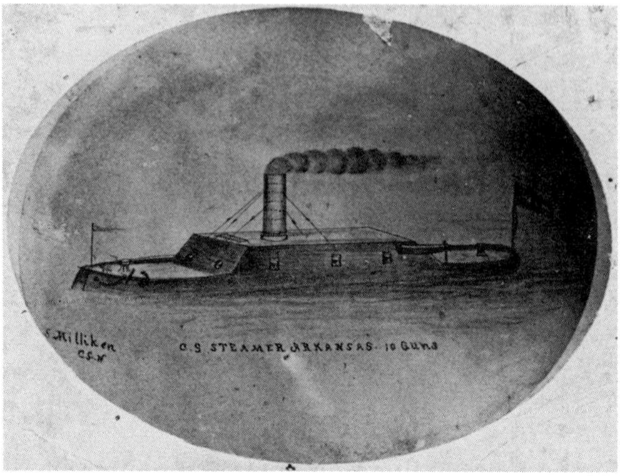

Figure 3.2. Contemporary sketch of the *Arkansas* by Acting Master Samuel Milliken, who served on board. Note the capstans on the fore and aft flush decks; on most other ironclads these were located inside the casemate.

The *Arkansas* and the *Tennessee* were designed to carry 2 propellers, each 6 feet 9 inches in diameter and driven independently by 2 horizontal direct-acting noncondensing engines that were each 30 inches in diameter with a 24-inch stroke. Four high-pressure boilers with 4,000 square feet of heating surface (fitted to burn coal) were provided for powering the engines at a maximum pressure of 75 to 90 PSI and a maximum speed of 75 to 80 RPM. Improvements were made during construction to the power plant by adding 2 more boilers, for a total of 6. The engine stroke was also increased to 30 inches, and the screw diameter to 8 feet.[42]

The *Arkansas*'s horizontal engines were typical of their type, being low down in the hull for protection and accessibility, but their weakness, according to Commander Brown, was a short piston stroke, which caused stoppages on top and bottom dead centers. Although no plans exist for the *Arkansas*'s engines, they were very similar to the preserved engines of the gunboat CSS *Chattahoochee* in Columbus, Georgia. The engine cylinders were even about the same size.

The completed *Arkansas* was 165 feet in length between perpendiculars, 35 feet in molded beam, and 12 feet in depth of hold. The design was unusual because the casemate had vertical sides, and the fore and aft flush decks were equipped with low-armored bulwarks.[43] Below the waterline the hull was deep and rounded like that of an ocean vessel; Commander Brown wrote that the ship "appeared as if a small seagoing vessel had been cut down to the water's edge at both ends, leaving a box for guns amidships."[44] Several contemporary sketches by other crew members corroborate this apt description (fig. 3.2).

Third Assistant Engineer E. A. Jack, who served on CSS *Virginia* and provided a valuable account in his memoirs, was assigned to the engineering staff of the *Arkansas* after the *Virginia*'s destruction. The change from Chesapeake tidewater to backwoods Mississippi must have been difficult, but as always, Jack (at least, according to himself) jumped right into his new duties. His first impression of the compartments he would be living and working in were not encouraging. Describing conditions in the engine room, he wrote, "The quarters were cramped, hot, and badly ventilated. In the engine room, which was partitioned from the ward-room access only by a rough board fencing about four or five feet high, the thermometer indicated over 100. Such a temperature when the engines were at rest and the fires low gave promise of trying and exhausting duty when underway or in action."[45] He would not have to wait long to find that these evaluations were only too accurate.

Commander Brown was soon ready to take the fight to the enemy after hurriedly finishing the ironclad and scrounging together a crew. He had heard of a Union fleet massing near Vicksburg, and on the morning of July 14, 1862, CSS *Arkansas* departed for battle. Much is written about the subsequent events; only the engineering highlights will be recounted here.

Before the new ironclad had gone very far, the hasty construction of the *Arkansas*'s steam plant became apparent to all. According to Commander Brown, "Fifteen miles below [Satartia, Mississippi], at the mouth of the Sunflower River, we found that the steam from our imperfect engines and boilers had penetrated our forward magazine and wet our powder so to render it unfit for use."[46] The crew was able to dry the powder by laying it in the sun, but valuable time was lost.

Nothing happened despite the captain's worries until dawn the next day, July 15, as the *Arkansas* approached the confluence of the Yazoo and Mississippi Rivers. Three enemy vessels on reconnaissance patrol were encountered there: the ironclad USS *Carondelet* (commanded by Brown's old friend Henry Walke), the wooden gunboat *Tyler*, and the ram *Queen of the West*. The Confederates gave chase and exchanged fire in a sharp battle as the Union vessels retreated toward the Mississippi. Although the *Arkansas* incurred some damage, the Union received the worst of it. The *Carondelet* was disabled and run aground, and the *Tyler* managed to escape with severe damage. Brown continued:

> On gaining the Mississippi, we saw no vessels but the two we had driven before us. While following these in the direction of Vicksburg I had the opportunity of inspecting engine and fire rooms, where I found engineers and firemen had been suffering under a tempera-

> ture of 120° to 130°. The executive officer [First Lieutenant Henry K. Stevens], while attending to every other duty during the recent firing, had organized a relief party from the men at the guns, who went down into the fire-room every fifteen minutes, the others coming up or being, in many instances, hauled up, exhausted in that time; in this way, by great care, steam was kept to service gauge, but in the conflict below the fire department broke down. The connection between furnaces and smoke-stack (technically called the breechings) were in this second conflict shot away, destroying the draught and letting flames come out into the shield, raising the temperature there to 120°, while it had already risen to 130° in the fire-room.[47]

The "second conflict" that Brown referred to was the arrival of the *Arkansas* at Vicksburg and the engagement between that vessel and no less than the combined fleets of Union Flag Officers David Farragut and John Porter (the former having ascended the Mississippi from New Orleans, and the latter having descended from upriver). In the meantime, engineer Jack corroborated Brown's account of conditions in the engine and boiler rooms during the battle: "The heat in the engine room was almost unbearable and in the fire-room the glowing red furnace doors shot out such heat that the men had to be frequently relieved and would gladly take a turn to rest on the quarterdeck where shot were flying around in preference to staying below where they were better protected. I do not remember the temperatures then, but do not think it extravagant to say that it was above 130."[48]

CSS *Arkansas*'s moment of glory came when Commander Brown led his already battered ironclad through the combined Union fleet off Vicksburg, singlehandedly engaging more than fifty enemy vessels. The Confederates inflicted great damage and confusion but barely made it to safety under the guns of Vicksburg itself. "It has been asked why the Arkansas was not used as a ram," Brown recalled. "The want of speed and of confidence in the engines answers the question. We went into action in Old River [the Yazoo-Mississippi confluence] with 120 pounds of steam, and though every effort was made to keep it up, we came out with but 20 pounds, hardly enough to turn the engines."[49] The loss of steam was a result of the tremendous punishment the *Arkansas* took from enemy shot: the unarmored stack was so perforated (and the firemen below so exhausted from the heat) at the end of the sortie that there was barely any draft for the fires.

Second Lieutenant George W. Gift (born in 1833), a Tennessean who lived in California before returning home to join the Confederate navy, provided a detailed account of his service on the *Arkansas*. Gift served on several vessels (including the aforementioned CSS *Chattahoochee*) in North Carolina, Georgia, Florida, and Alabama. His experience on the *Arkansas* commanding the forward port battery (one forward and one port firing gun) was certainly a high point for the young officer.[50]

Gift further described the scene on board during the Vicksburg battle: "But still the ship was not disabled; seven guns [out of ten] were still hammering away, and the engines were intact. But steam was down to a terribly low ebb. The party who had fitted up the boilers had neglected to line the fire front with non-conducting material; the consequence was that when a heavy fire of coal was put in the whole mass of iron about the boilers became red-hot and nearly roasted the firemen, who had also got a tub of ice-water, of which they drank freely."[51] The battered *Arkansas* finally reached the wharves of Vicksburg under the city's protective guns on the bluffs, having scored one of the most dramatic victories of any Confederate ironclad.

The *Arkansas* sat at Vicksburg for a week receiving repairs but was constantly under bombardment from the Union fleet. Farragut's forces steamed to a point below the batteries at Vicksburg during the night of July 15 and hammered at the Confederates as they passed. Serious damage was inflicted in the engine room. Gift wrote, "An eleven-inch shot pierced our sides a few inches above the water-line, and passed through the engine-room, killing two men outright (cutting them both in two) and wounding six or eight others. The medicines of the ship were dashed into the engine-room, and the debris from the bulkheads and splinters from the side enveloped the machinery."[52] This shot may have partly accounted for the increasingly temperamental state of the *Arkansas*'s power plant on its final voyage. It apparently did significant damage to one of the engines. In the meantime, further conflicts with the Union ironclad gunboat *Essex* continued to hamper the repair efforts. This large warship later became the *Arkansas*'s particular nemesis. Finally the Union fleets retired to the south (Farragut was worried about his deep-draft seagoing vessels becoming stranded by falling river levels) after a week of ineffectual attempts to destroy the Confederate ironclad, and Vicksburg was temporarily left in peace.

The *Arkansas*'s final voyage came in August when it was ordered south to aid in an attempt to retake Baton Rouge, Louisiana. On August 3 Commander Brown and Chief Engineer George W. City were away on sick leave, and the ship was placed in charge of the executive officer, First Lieutenant Henry K. Stevens.[53] Consequently, the *Arkansas* left Vicksburg for battle without the two men most crucial to its smooth operation.

The temporary commander was certainly a capable officer, but he was no Isaac Newton Brown. Stevens was born in South Carolina in 1824 and served in the US Navy before joining the new Confederate sea service in November 1861. He was posted to Memphis shortly before Commander Brown and served with him at Yazoo City after the *Arkansas*'s destruction. He later assumed command of the partly armored tinclad gunboat CSS *J. A. Cotton* in Louisiana. His promising career as a Confederate captain was cut short when he was killed in action on the *Cotton* near Patterson, Louisiana, on January 14, 1863.[54]

During his time in command of the *Arkansas* Lieutenant Stevens tried to follow Brown's example in bold leadership, but Chief Engineer City's absence in particular began to trouble him. Many problems in the engineering department soon developed. Second Lieutenant Gift stated the following:

> We left Vicksburg thirty hours before General Breckinridge [the Confederate commander at Baton Rouge] had arranged to make his attacks. The short time allowed to arrive at the rendezvous made it imperative that the vessel should be driven up to her best speed. This resulted in the frequent disarrangements of the machinery and consequent stoppages to key up and make repairs. Every delay required more speed thereafter in order to meet our appointment. Another matter operated against us. We had been compelled to leave behind, in the hospital, our chief engineer, George W. City, who was worn out and broken down by excessive watching and anxiety. His care and nursing had kept the machinery in order up to the time of leaving. We soon began to feel his loss. The engineer in charge, a volunteer from the army, had recently joined us, and though a young man of pluck and gallantry, and possessed of great will and determination to make the engines work, yet he was unequal to the task. He had never had anything to do with a screw vessel or short-stroke engines, and, being zealous for the good repute of his department, drove the machinery beyond its powers of endurance.[55]

Gift's statement reveals the disparity between naval- and civilian-trained engineers; in the western theater the latter worked on paddlewheel-driven riverboats and were generally unfamiliar with propeller technology. Civilian steamboat men also usually had no formal training and chafed under navy regulations and supervision. These are the major reasons that formally trained men like City were sorely needed for new warships like the *Arkansas*.

George Washington City was another prime example of the young but highly skilled set of steam engineers who were often far more familiar with the latest technological developments than the older deck officers. He was born in Washington DC in 1835 and served in the US Navy as a first assistant engineer from 1854 until his resignation in July 1861. City's rank was transferred to Confederate service, and he was assigned to the training ship CSS *Patrick Henry* on the James River before joining the crew of CSS *Virginia*. Like E. A. Jack, after that ship's destruction he found himself transferred to the bayous of the Mississippi and duty on another ironclad. After the *Arkansas*'s loss, City served at shore stations in Mississippi and Georgia and also on several different vessels, including CSS *Chattahoochee*. He finished the war in Augusta, Georgia, and afterward lived in Baltimore until his death at an unknown date (after 1900).[56]

As the *Arkansas* began its last cruise, engineer Jack was in the engine room and provided firsthand testimony of the developing problems. His detailed account suggests that it was unnecessary adjustments made to the engines, not simply faulty manufacture or hard use, that ultimately ruined them:

> Indeed the engines were good as they were, and as capable of carrying the vessel through an attack on the enemy as they had ever been. This new Chief [the army volunteer] . . . wanted to make alterations in the lap of the steam valve [*lap* is the period when both admission and exhaust valves to the cylinder are closed to allow for expansion of the steam in the cylinder]. . . . I had just examined the set of it and was closing the steam chest when I became aware of the presence of a stranger who was looking on. "I want to look at that valve," he said. I regretted that he had not come aboard sooner as it was too much trouble to open the chest again and I was very busy. But he replied, "I have come on board by Capt. Brown's order to take charge of these engines." I did not know whether I was talking with a lunatic or not, but Mr. Stevens, coming up at this time, confirmed the newcomer's rank and I proceeded to obey orders. I took off the cover and disconnected the valve and took it from the chest. Then it was that he sagely remarked, "I do not wonder that this thing does not work: it is an inch too long on both ends. I will make it shorter." I cannot express my surprise when he said this, for I knew that if he did it, the valve would be ruined and the vessel delayed until a new one could be gotten. How long that would have been I cannot say, but I doubt if one could be gotten short of Richmond. The old valve might have been repaired, but that too would have taken some time. I could not help telling this new and *experienced* officer that he would ruin the valve if he altered it as he contemplated, but he smiled knowingly, and, as if it was presumption in a youngster to differ with him, turned away. I protested then to Lieut. Stevens, but he said, "This officer is more experienced than you

> and I suspect he knows what he is talking about." I then asked that he would speak to Mr. City. Mr. City tells me that he did, and that he told Mr. Stevens that I was right and the man was a fool. But he was kept although the alterations were not made in the valve.[57]

Jack was more knowledgeable of the *Arkansas*'s engines, so Stevens deferred to the other man's experience even though it was not applicable in this case. He was simply following hierarchy protocol, but the encounter also shows Stevens's ignorance (possibly willful) of steam machinery operation. This was common in deck officers—to this day there is a distinct dichotomy if not rivalry between officers and engineers on naval vessels. Jack seemed to have helped prevent a serious problem, but ultimately the engines were tampered with during what should have been routine maintenance. Jack's continuing detailed account makes abundantly clear that he believed such blunders were the cause of the *Arkansas*'s downfall:

> Lieuts. Stevens and Gift have both written that it was due to a breakdown of the engines. The loss of the vessel was indisputably due to that, but what led to that breakdown has never been written. It might reasonably be asked why the engines that had brought this ship through the fleet off the Yazoo River, fighting every inch of the way, and which had been so successfully handled when the two attacks were made upon her at Vicksburg, had become so inefficient at this critical time. A practical man would suspect that they had been badly managed, indeed tampered with, and would not be very far from the truth. Our trip was one of accidents and delay from the beginning to the end.[58]

Jack related valuable technical detail from his firsthand involvement in what happened next:

> The first stop because of the machinery was unavoidable. It was caused by a loose key in the rock-shaft arm [part of the valve, or reversing, gear]. An examination disclosed that the key-way was very much worn, and it was thought advisable to true up the way and fit a new key. The link was disconnected, the repairs made, and the machinery coupled up again. This latter duty was carefully performed under the supervision of the experienced chief engineer [William M.] Fauntleroy, and I did the work on the rock-shaft, and when that was finished, took a rest which we deserved. The others [repaired pieces] our relief connected up. In fact I am quite certain that they disconnected the

> link. At any rate, no marks were made by which a proper length of the eccentric rods might be preserved, and in putting them up again one was left longer than normal and when the engine was moved it drove the steam valve against the end of the chest and broke the pin in the rocker-arm which had just been secured. Our work held, but the pin gave way. A new pin was forged, and Fauntleroy fitted that and gave it to an assistant named Gettis to secure, but he took it and heated it red hot to make it easier to rivet in place, and tried to drive it into place hot. Of course it would not fit as it had expanded from the red heat it had, but he up [*sic*] with sledge and tried to drive it home. In this he failed, but he succeeded in spoiling the pin which had to be fitted again.[59]

Jack and the others became more exasperated as the voyage continued. The engines grew more temperamental as the *Arkansas* hurried south. Lieutenant Stevens began to be worried about taking his ship into action and whether he would even be able to make it to Baton Rouge.

Second Lieutenant Gift continued his account of the *Arkansas*'s last voyage: "At or near the mouth of Red River, the engines had grown so contrary and required to be hammered so much that Stevens deemed it was his duty to call a council of war to determine whether it was proper to proceed or return. The engineer was summoned and gave it as his opinion that the machinery would hold out, and upon that statement we determined to go ahead. A few miles below Port Hudson he demanded a stoppage to key up and make all things secure before going into action."[60] The *Arkansas* continued with good speed toward Baton Rouge after this last repair, but by the time the ironclad reached the city, the land battle was all but over. Confederate General John C. Breckinridge had lost the element of surprise, and the opposing Union forces were helped to victory by supporting warships on the adjacent river, something the Confederates did not have because of the *Arkansas*'s late start and increasingly critical machinery problems.

Jack spelled out his views about the culpability of the civilian engineering specialists and the frantic work belowdecks before the final battle:

> When I went on watch I was his [civilian assistant Gettis's] relief, and going around with him to examine into the condition of the department, I saw a machine I was not familiar with. I knew it to be a Giffard's injector, but [did] not know how to operate it. So I asked him to show me how it started. Much to my surprise he declined to do so, and told me to learn to operate it as he had. I had the authority to compel him to show me, but my professional pride was touched, so throw-

> ing aside our official relations, I determined to show him that I could do as he had so uncivilly advised. And when he had gone, after a few trials, I succeeded in manipulating the machine successfully. Well, after the new pin [referred to previously] was fastened in place and the eccentrics adjusted, another start down the river was made. But next the coal supply was found to be insufficient and another landing was made at a plantation to replenish the bunkers. Whether this was an oversight or there were no coals in Vicksburg, I don't remember. While the coaling was going on, the starboard cylinder was opened and the packing in the piston was set out. [Packing was sealing material for the cylinder. It was commonly made of hemp.] This engine worked the port wheel, which you will learn hereafter became inoperative and caused the loss of the vessel. This tinkering with the engine was altogether unnecessary. The civilian engineer was, in my opinion, culpable in permitting it, and guilty as well of negligence in not seeing that the packing was properly adjusted, instead of entrusting such an important matter to an assistant who was not familiar with it. . . . Day came before we had finished coaling, and when it was quite light we cast off our lines and started down river once more.[61]

At a clear area above Baton Rouge, the crew learned that USS *Essex* and other Union vessels were still present. The Confederates decided to attack them. Gift wrote the following:

> Stevens assented to the proposal [to ram the *Essex*] and had just remarked that we had better go to our stations, for we were in a hundred yards of the turn [in the river], when the starboard engine stopped suddenly, and, before the man at the wheel could meet her with the helm, the ship ran hard and fast aground, jamming herself on to some old cypress stumps that were submerged. . . . On investigation it was found that the engine was so badly out of order that several hours must be consumed before we could again expect to move. There lay the enemy in plain view, and we as helpless as a shear-hulk.[62]

Commander Stevens saved a copy of the frantically prepared engineering report submitted to him at this time. Its main point was this: "From my examination of all the defects in the . . . engines, it will be impossible for me to put them in fit condition to render efficient aid to propel the ship into action."[63] The *Arkansas* now appeared to be in dire straits. The end was soon in coming. Second Lieutenant Gift described the final calamity as follows:

> Stevens was not the man to give up. A quantity of railroad iron, which had been laid on deck loose, was thrown overboard, and in a few hours we were afloat. The engineers had pulled the engine to pieces and with files and chisels were as busy as bees, though they had been up constantly then for the greater parts of the two preceding nights. At dark [on August 5] it was reported to the commanding officer that the vessel could be moved. In the meantime some coal had been secured (our supply was getting short), and it was determined to run up stream a few hundred yards and take it in during the night, and be ready for hot work in the morning. Therefore we started to move, but had not gone a hundred yards before the same engine broke down again; the crank pin (called a "wrist" by western engineers) of the rock-shaft broke in two. Fortunately one of the engineers was a blacksmith, so the forge was set up and another pin forged. But this with our improvised facilities used up the whole night. Meantime the enemy became aware of our crippled condition, and at daylight moved up to the attack. The *Essex* led, and came up very slowly, at a rate not to exceed two miles an hour. She had opened on us before the last touch had been given to the pin, but it was finished and the parts thrown together.... The pleasant sensation of again being afloat and in possession of the power of locomotion was hardly experienced before our last and final disaster came. The port engine this time gave way, broke down and would not move. The engineer was now in despair; he could do nothing, and so reported. The *Essex* was coming up astern and firing upon us. We had run ashore and were a hopeless, immovable mass.[64]

As USS *Essex* once again approached CSS *Arkansas* to finish the business begun at Vicksburg, its shots began to fall around the stricken Confederate ironclad. The gun crews of the latter were quick to respond. In the midst of the deafening reports, confusion reigned below in the engineering spaces. The crews worked desperately to get the *Arkansas* moving again. Engineer Jack's account of those last moments adds further harrowing detail:

> Immediately after our guns opened fire [on the *Essex*], one of our engines stopped. My impression then was that the shot that had struck so close astern of us had struck our propellor [*sic*], for its fall, the discharge of our guns, and the stoppage of our engine came very close together. I rushed down from the casemate and passed below through the hatch in the overhang aft, and just as I had gotten below,

Figure 3.3. Contemporary drawing of the *Georgia*. The configuration of the casemate seems to agree with other contemporary sources and recent archaeological findings, although the lack of a smokestack is puzzling.

> another of our after guns was fired. I found them trying to move the starboard engine, which was somehow jammed and the engine room full of steam that was escaping from the overstrained joints. I felt sure then that our shaft had been bent by the shot which had struck astern of us, for I did not know that the packing had been set out in that engine as I was ashore when it was done, attending to the coaling. (I learned this afterwards from Asst. Engineer Fauntleroy who saw it done.) If I had known it, there would have been no doubt in my mind as to the cause of the engine stoppage. All efforts to start the disabled engine failed. The port engine was kept at work and in a short time the ship was run ashore on the west side of the river, and abandoned and fired. . . . The excitement in the engine room because of the disabled engine and the pell-mell way in which both officers and men abandoned the ship . . . completely unnerved me for a while.[65]

It is interesting that Jack made no mention of the port engine breaking down. This difference in detail may simply be a result of Gift's not actually having been in the engine room during the final battle. He had enough to do in leading his gun crew and may have misremembered the final sequence of events. In the end, it did not matter. The *Arkansas*'s machinery had stopped for the last time.

In the face of enemy fire and risking capture by Union troops moving up the bank, Lieutenant Stevens made the painful decision to destroy his vessel. Accordingly, the crew abandoned ship into the swamp, and the Confederacy's most gallant ironclad burned and blew up. According to historian H. Allen Gosnell, "Her career lasted only twenty-three days, but what a career! It included so much action that there probably never was another vessel that averaged anything like as much fighting per day as did the *Arkansas*."[66]

The engines of the *Arkansas* were ultimately its greatest weakness, despite the apparent superiority of being purpose built coupled with the advantages of independently operating screws. Repeated breakdowns in the face of the enemy made the ship's destruction by its own crew necessary. That crew, meanwhile, performed its duties to the utmost, and several crew members had the foresight to record their experiences and descriptions of the ironclad. This is fortunate, because no plans, photographs, or machinery schematics are known to survive. A future archaeological survey of the *Arkansas*'s wreck near Baton Rouge may reveal more details. Of particular interest would be to investigate whether the engines were modified from riverboat machinery components, given the manufacturing location.[67]

CSS *Georgia*

CSS *Georgia* (sometimes called the *State of Georgia* or the Ladies' Gunboat), a self-powered floating battery constructed in Savannah, remains to this day one of the most enigmatic Confederate ironclads despite extensive surveys of its wreck by the US Army Corps of Engineers. Like the *Louisiana* and the *Arkansas*, the *Georgia* was a large early example of Confederate ironclad construction hampered by a faulty propulsion system. The ship ultimately served out its existence as a floating battery stationed near Fort Jackson below Savannah and was scuttled when that city was captured by William T. Sherman's Union forces on December 20, 1864.

Almost nothing is known of the *Georgia*'s hull construction other than that it must have been simple—building time was only two months.[68] In 2015 the wreck was raised from the bottom of the Savannah River and extensive archaeological documentation was done, but nothing remains of the lower hull for a totally accurate reconstruction to be made.[69]

The *Georgia*'s design is attributed to Alvin N. Miller, a local foundryman with little experience in naval architecture.[70] Miller, born in New York City in 1814, was an experienced machinist who was involved with numerous shipbuilding and engineering projects since his youth. He moved to Savannah in the early 1830s and established himself as a prominent mechanic, businessman, and well-connected local citizen, although he often accumulated large debts. After completing the *Georgia*, Miller's foundry produced more ordinary products for the navy such as screws, rivets, bolts, washers, and cut timber.[71]

The *Georgia* was financed by the Ladies' Gunboat Committee of Savannah and laid down sometime in March 1862, probably at Harding's Shipyard, adjacent to Miller's foundry.[72] Although information on the ironclad's dimensions exists, it is contradictory. Stated measurements vary from 150 feet long and 50 feet in beam to 250 feet long and 60 feet in beam.[73] Several contemporary drawings of uncertain accuracy depict the above-water portion of the vessel (fig. 3.3), but no plans of the *Georgia* are known to survive.

After its launch on May 19, 1862, the *Georgia* proved a disappointment. Although it was well-known that the ironclad was constructed only as a self-powered floating battery and not as a gunboat, as many had wished, a trial trip revealed that the *Georgia* could barely move and that it leaked quite badly. Attempts to gloss over these defects were largely unsuccessful, but the ship still answered the purpose of a floating defense. It was accordingly towed down to a specially prepared mooring place and guarded the river approaches to Savannah. The *Georgia* remained in that spot for the rest of its service life.[74]

Little information exists on the ship's machinery other than that uncovered by archaeological investigation. All that was known before this was that the *Georgia* had been completed as a twin-screw vessel with double engines.[75] They were not able to move the vessel at any more than two knots even with the current, but they did serve the important purpose of keeping the pumps running for combating numerous leaks caused by the use of green wood in the hull.[76]

The exact origin of the *Georgia*'s power plant is not clear. The floating battery may have been powered by machinery built or supplied by Miller because of its construction location next to his foundry and machine shop. There are also other possibilities. Historical records dating from the time of construction reveal that the builder's committee searched for engines far and wide.[77] It has been suggested that the engines came from the sidewheel steamer *William Jenkins*, but since the engines have now been recovered and fully documented, it is obvious that they do not represent a beam engine like that on the *William Jenkins*. The engines are actually of the standard horizontal type, suggesting they may have been purpose built. Miller's foundry certainly had the capability of producing these engines, because he cast cylinders for a steam locomotive in the 1850s and manufactured at least one heavy weapon during the Civil War.[78] Apparently the foundry also operated as a local outlet for the Columbus Naval Iron Works.[79]

Another cited possibility is the machinery of the steamer *Emma*. All that is known about this vessel is that it was a river steamer, with two disconnected engines, eventually run aground and burned to keep it out of enemy hands. No other substantial technical information on this ship has been found.[80]

Firsthand accounts of the *Georgia*'s uneventful wartime service are fairly scarce, and no engineering details beyond the vague double-engine reference have been located. Service on board the ship was not pleasant: constant heat from the machinery (run regularly to keep the pumps running; the vessel's lack of movement would have denied the crew any cooling breeze), dampness from leaks, and sheer boredom would have frustrated even the most unmotivated sailor. Fortunately, diary entries describing a similar situation on the *Richmond* class ironclad CSS *North Carolina* exist, providing a glimpse into what daily life must have been like on the *Georgia* as well. Those writings (made by a junior engineer) will be discussed later in connection with the *North Carolina*.

The *Georgia*'s captains over the course of its career—Lieutenants John Pembroke Jones, Thomas P. Pelot, Charles F. M. Spotswood, and Washington Gwathmey—all had to combat the problems of running what was essentially just a floating guard post. Jones later moved on to command the ill-fated *Richmond* class ironclad CSS *Raleigh*, whereas Pelot and Spotswood were temporary captains. Gwathmey was an Englishman by birth but sided with the Confederacy after living in Virginia. His experience in prewar service with the US Navy transferred over to Confederate service, and Gwathmey commanded vessels at New Orleans and saw shore duty in Richmond and Fort Caswell, North Carolina, before being transferred to the Savannah station.[81]

Gwathmey was stuck on the *Georgia* for the rest of the war after he became captain on July 10, 1863. His primary duties during that time were quarterly target practice (probably the highlight of duty on the *Georgia*) and trying to make life on board bearable.[82]

Despite the miserable experiences of those on board and the myriad defects in the *Georgia*'s construction and machinery, the ironclad proved useful enough as a floating defense that when Sherman's forces moved on Savannah in late December 1864, at least one attempt was made to tow the floating battery to a new position. When this failed, the crew was evacuated on the night of December 20 and the *Georgia* was scuttled. This method of sinking accounts for the amount of preservation at the wreck site. Virtually every other Confederate ironclad was blown up at the end of its career.[83]

The *Georgia* was ultimately one of the most defective Confederate ironclads, but it played an important role in the defense of Savannah. No ironclad in the Savannah Squadron served as long or proved as useful. Hastily built and never used in anger, it nonetheless provided security along the river and helped keep the enemy at bay for three years.[84] Any problems with the ship's design were really only secondary to such a vital purpose.

If detailed knowledge of the *Georgia*'s machinery, its greatest defect, existed among anyone other than those who built the vessel, it was lost over the years. In the twentieth century, archaeology finally began to provide some answers about this obscure warship's design, construction, and propulsion systems. The methods that were employed in investigating, documenting, and raising the *Georgia*'s wreck could possibly be used on future excavations or examinations of other Confederate ironclad wrecks.

The *Georgia*'s steam machinery remains remarkably intact. During the course of archaeological investigations, elements of the engines, boiler(s), and propulsion were discovered and later raised along with the remainder of the vessel's structure. All the findings corroborate the description of a double engine given when the ironclad was afloat. It appears that the *Georgia*'s two engines were the single-cylinder horizontal direct-acting type, each with a diameter of 24 inches, a length of 39.5 inches, and a piston rod diameter of 3 inches. In addition, a third cylinder was found that may have been a condenser. This piece of machinery measures 9 feet 4 inches in length and 20 inches in diameter, with attached piping and valve components.[85]

A number of disarticulated iron boiler plates, staves, rods, and firebricks were found in addition to the engine and condenser cylinders. The layout of these remains suggests a single rectangular boiler about 18 feet long, 7.5 feet wide, and very similar in design to the marine variant of locomotive (firetube) boilers commonly used on small craft during the nineteenth century. It is unknown whether the *Georgia* had more than one boiler.[86]

One of the propellers and a portion of shafting were also found. The three-bladed screw is 9 feet in diameter, the hub is 18 inches in diameter and 12.5 feet long, and the attached shaft fragment is six inches in diameter. A stuffing gland and bearing are also present, denoting the point at which the shaft passed through the hull.[87]

As of 2016 the raising of all artifacts from the wreck site had been completed as part of a collaborative project between the US Navy, the US Army Corps of Engineers, and institutions and individuals involved in cultural resource management. Removal of the wreck was deemed necessary in order to deepen the approaches to the port of Savannah for modern container ships. Artifacts are now undergoing conservation and cataloguing at Texas A&M University in College Station.

Conclusion

The three completed early-war nonstandard ironclads represented various configurations like the converted vessels before them but were all compromised by an uncertain operating philosophy. That philosophy (primarily Secretary Mallory's) intended the *Louisiana* and the *Arkansas*, at least, to be combination defensive-offensive units capable of both inshore and coastal work. Accordingly, these vessels (as well as the *Georgia*) were uncommonly large ironclads by later standards. Because of this the Confederacy lost valuable time and resources in constructing them, instead of building several smaller vessels better suited for a specific harbor defense role like the *Richmond* class ironclads. The compromises made in the *Louisiana*'s, the *Arkansas*'s, and the *Georgia*'s designs greatly affected their performance characteristics. Only the *Arkansas* can be considered a success as a self-propelled warship.

The primary problems with the nonstandard ironclad designs were related to their machinery installation. This fact is more surprising than that related to the previous ironclad conversions, because according to the available facts, the boilers, engines, and propulsion systems for the *Arkansas* and the *Georgia* were purpose built. The *Louisiana*'s components were suitable for their original vessels but could not move such an unwieldy construction. Thus the *Louisiana*'s mobility problems primarily resulted from its unconventional layout rather than the machinery used.

The skill shown by Southern manufacturing establishments in building the power plants of these three ironclads, again like the converted vessels, reflected the relative youth of steam engineering as a science (especially in the agricultural Confederacy) coupled with the troubles inherent in designing systems for such revolutionary and experimental vessels. This lack of experience was also shown in operation: in the *Arkansas*'s case, at least, the naval engineers and civil appointees butted heads over proper handling, and the ship's loss may have been a direct result.

All in all, it appears that the engineers on board the *Louisiana*, the *Arkansas*, and the *Georgia* did the best they could with what they had. Although these ironclads and their machinery certainly represented a next step in Confederate armored-warship design, that step was largely unsuccessful. What the Confederacy really needed was a standardized ironclad model.

FOUR

THE *RICHMOND* CLASS IRONCLADS

We ought to construct several like her [the *Richmond*].
—Lieutenant John M. Brooke, Confederate States Navy

The Confederacy's first class of standardized ironclads resulted from a solidification of ideas on how best to employ armored warships. Whereas the first ironclads such as CSS *Virginia* were intended for mixed defensive-offensive operation, by late 1861 to early 1862 the Confederate Navy Department realized it was unable to construct large seagoing ironclad warships capable of breaking the Union blockade.[1] A defensive strategy to protect the primary ports (still crucial for the delivery of necessary goods by blockade-runners) was therefore adopted. To fit this purpose a number of relatively small ironclads of simplified design were planned by Naval Constructor John Porter. Six of these vessels were ultimately completed, never formally grouped by the Confederates but known today as the *Richmond* class.

Porter's standard hull design for the *Richmond* class ironclads (the *Richmond*, the *Chicora*, the *Palmetto State*, the *North Carolina*, the *Raleigh*, and the *Savannah*) was also used on several subsequent ironclad classes. For the most part it served credibly during the war.[2] The *Richmond* class vessels set this precedent.

Though certainly an improvement over the earlier nonstandard designs like CSS *Louisiana*, these new ironclads still had some noticeable flaws. Most prominent was a draft too deep for the majority of southern harbors.[3] Also, like virtually all Confederate ironclads, CSS *Richmond* and most of its sisters suffered from inadequate machinery.

CSS *Richmond*

The prototype for the largest single class of ironclad completed by the Confederacy was the Norfolk-built CSS *Richmond*. The design was largely based on Constructor Porter's proposed 1846 Pittsburgh model.[4] Current research indicates that more than twenty ironclads eventually incorporated the *Richmond*'s design, although only the six discussed here were ever finished and placed in service.[5] The *Richmond* was the first completed and naturally set the trend for all the following ships, although there were many noticeable differences between units.

Construction on the *Richmond* began at the Gosport Navy Yard in Norfolk before that facility's abandonment by Confederate forces in May 1862.[6] Consequently, the unfinished ironclad was hurriedly launched on May 6 and towed up the James River to the capital.

Several of the *Richmond* class ironclads received their power plants from other vessels. A continuing lack of suitable engines led to scavenged machinery for four of them, beginning with the *Richmond*. Its steam machinery was particularly weak and reportedly removed from a former lightship with a varied service life, USS *Arctic*.[7]

The history of the *Arctic* is interesting if somewhat murky. This is largely because of an error in the records that unaccountably confused the *Arctic* with the paddlewheeler *Thomas G. Haight*.[8] What is known is that the *Arctic* was built at the Philadelphia Navy Yard in 1855 and worked for a time as a US Coast Survey ship exploring northern waters and along the Eastern Seaboard. The *Arctic* was brig rigged but carried steam machinery and a single screw. The manufacturer was Franklin Iron Works, James T. Sutton Company of Philadelphia.

The *Arctic*'s hull measured 98 feet long between perpendiculars, 23.5 feet in molded beam, and 11 feet in depth of hold, with a 232-ton displacement.[9] Nothing is known about its engine other than that it was probably a single-cylinder vertical-inverted direct-acting type like that of most tugboats and small screw steamers.

By the time of the Civil War, the *Arctic* was serving at Frying Pan Shoals in North Carolina as Lightship No. 8. It had served in this station since 1859. Original plans and line drawings of the vessel show its configuration as a typical lightship of the time (fig. 4.1).[10] During the conversion process into a lightship, the original machinery was removed and placed in storage. It was apparently located at the Gosport Navy Yard when obtained by the Confederates.[11]

Lightship No. 8 in turn fell into Confederate hands when it was sent to Wilmington, North Carolina, in April 1861 for a routine overhaul. The vessel was seized and renamed CSS *Arctic*. It remained on the Cape Fear River throughout the Civil War as a receiving ship and ironclad floating battery.[12]

Meanwhile the recipient of the *Arctic*'s engine was finished. The *Richmond* conformed closely to Porter's original design with a length between perpendiculars of 150 feet, a beam of 34 feet, and a draft of 13 feet.[13] The overall length was approximately 174 feet, with a maximum beam over the ship's built-on armored knuckle of 44 feet.[14] Although there are very

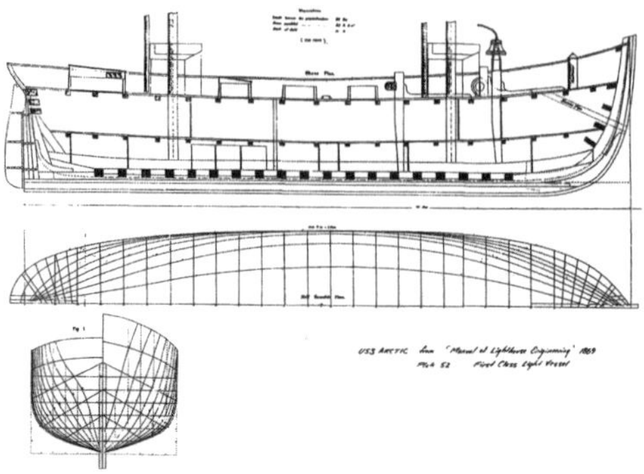

Figure 4.1. Lines of USS *Arctic* as Lightship No. 8. The propulsion machinery is removed and the standard lightship equipment installed. This includes a large central bow hawsepipe for the mushroom anchor that holds the ship on station, an oversized windlass on the foredeck for controlling the anchor gear, a fog bell on the bow, lamp houses around each lantern mast for servicing the beacon chandeliers, and smaller spencer masts behind each main mast for carrying steadying sails.

few contemporary images of the *Richmond* (and those show it only from a distance), detailed original plans of a sister ship, CSS *Savannah*, record its general appearance (fig. 4.2). It can therefore be safely stated that like its following class members, the *Richmond* had short flush decks identically formed at bow and stern, a long central casemate, four main guns, and a shiplike lower hull propelled by a single screw 8–10 feet in diameter.

Steam was supplied by two brand-new horizontal low-pressure fire-tube marine boilers built by either Tredegar Iron Works or the Confederate Naval Works in Richmond.[15] The exact dimensions of the *Richmond*'s boilers are not known, but they probably conformed to the dimensions of the standardized type manufactured in Richmond and the Confederacy's other major industrial center, Columbus, Georgia: 10 feet long, 11 feet high, and 6 feet 9 inches in diameter. Standard tubular boilers saw service on many Confederate vessels, including the other *Richmond* class ironclads.

Although it is uncertain who actually built the *Richmond*'s boilers, Tredegar was definitely in charge of installing them. Those works were contracted to supply the machinery for the ironclads built at the Confederate capital, where Tredegar itself was located.

The mighty establishment of Tredegar Iron Works during the Civil War had a humble beginning dating back approximately thirty years. A Virginia blast-furnace operator, Francis Deane Jr., laid plans in 1836 to construct a new iron foundry. On a narrow plot of land in Richmond between the James River and Kanawha Canal, an engineer from the town of Tredegar in Wales designed the new facility. The new foundry, rolling mill, and forge were accordingly named Tredegar in his honor. Rail-

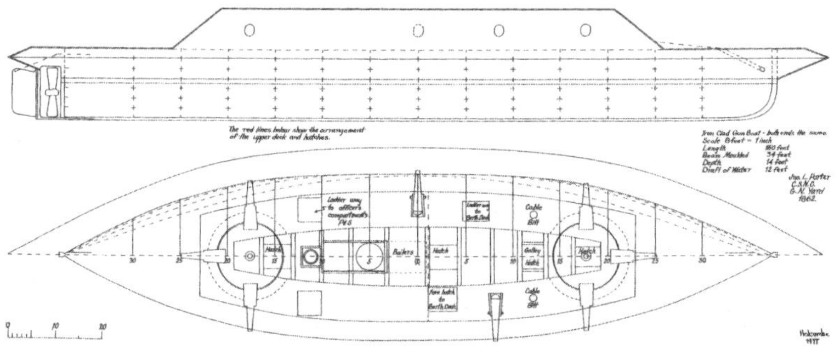

Figure 4.2. Plans of the *Richmond* class ironclad CSS *Savannah* redrawn by Bob Holcombe in 1978. The machinery is not shown, but these drawings give a fairly complete overview of the layout; the *Richmond* and others of the class were similar if not identical. Note the position of the smokestack, the engine room grating, and what appears to be a ventilator shaft aft of amidships on the top view plan. The location of the boilers is also marked.

road iron was the main product of the new company, which quickly fell on hard times after the 1837 economic crisis. Tredegar limped along until 1841, when it was sold to the young and ambitious Joseph Reid Anderson, a West Point graduate.[16] This man's incredible organizational and leadership abilities resulted in an industrial complex rivaled only by those in the industrialized Northern states.

The new owner of the ironworks quickly began to implement changes and improvements that lasted through the 1840s and 1850s. By 1861 Tredegar was a superb heavy industrial complex capable of turning out large quantities of ordnance, munitions, armor plate, and other desperately needed iron products, including steam engines.[17] Multiple blast furnaces, rolling mills, forges, and machine shops fueled the Confederate war effort. Anderson's facilities were the largest and most important industrial center in the Southern states until it was shut down in April 1865.

Although it is certain that Tredegar produced steam machinery during the Civil War, it is uncertain whether it built all the engine components for ironclads. Anderson contracted with the government to provide the plates, spikes, engines, shafts, and armament for the locally constructed ironclads.[18] The *Richmond* seems to have been fitted out by Tredegar, and two ironclads completed later, CSS *Fredericksburg* and CSS *Virginia II*, were also furnished by the works.

Tredegar may also have built or at least procured engines for other ironclads (possibly CSS *Raleigh* and CSS *Albemarle*) and in addition forged the massive central propeller shaft for CSS *Mississippi*. Another possibility is that building engines was outsourced to the Confederate Naval Works, commonly and formerly called the Shockoe Foundry, since the Tredegar contract only called for supplying engines. Therefore, the *Fredericksburg*'s and the *Virginia II*'s power plants may have been built at this other prominent establishment; the uncertainties pertaining to these ironclads' machinery will be discussed later.

So many questions exist because a great number of documents relating to naval construction were destroyed in the evacuation of Richmond on April 2–3, 1865. Little doubt remains that even if it did not construct whole engines, Joseph Anderson's Tredegar Iron Works contributed significantly to the installation of steam machinery components and other iron fittings in many Confederate ironclads.

The Shockoe Foundry in Richmond (formally called Talbott & Brother) was closely tied to Tredegar Iron Works in the manufacture of marine steam machinery. This facility produced small steam engines and sawmills before the war and was "considered very good here [Richmond] for what they purchase, and in all our [Tredegar Iron Works] transactions with them they have been very prompt."[19] This large establishment was important in the manufacture of steam machinery in the Confederate capital, but little is known of its true output because of the destruction of its records at the war's end.

By 1862 the Shockoe Foundry was more than a full city block in size. It was located at 17th and Cary Streets in Richmond. The large establishment consisted of a fitting shop, turning shop, smith's shop, foundry, drying oven, boiler room, and flash room.[20] By that time it was only surpassed in size and output by neighboring Tredegar, which Shockoe often supplied with parts and materials.

First founded in 1839 as the Shockoe Manufacturing Company by Charles Talbott, the former manager of the Seaboard and Roanoke Railroad, the foundry remained prosperous through the 1850s with the growth of Virginia's railroads. The firm even began producing its own locomotives in 1855.[21] By 1862 the Talbott brothers' excellent establishment earned the Shockoe Foundry a government contract to build engines for wooden gunboats. When that program fell through, power plants for ironclads were manufactured, but the extent of the new production is unknown.

In February 1862 the Confederate government stepped in and leased the Talbotts' establishment in the name of the navy as the Confederate Naval Works. Under the supervision of Thomas W. Smith, the foundry turned out wartime products for the rate of five thousand dollars per month.[22] It is thought that these works produced several engines for ironclads, but the dearth of records has led to confusion over whether the Naval Works or Tredegar produced the machinery. It is possible that Tredegar, which was tasked with providing all the iron components of the James River ironclads, contracted with the Naval Works to build the engines. These establishments had worked closely together before, and the capability to produce excellent quality parts of all types was certainly present.

Current research holds that the ironclads *Fredericksburg, Virginia II, Albemarle,* and *Raleigh* and several of the small wooden gunboats on the James River may have received all or part of their machinery from the Naval Works.[23] The *Richmond*'s new boilers may have been built there as well. In any event, it is clear that the former Shockoe Foundry played an important role in steam machinery construction in the Confederacy. It is hoped that this role can be more fully defined by future research.

Regardless of the *Richmond*'s power plant's origin, little could be done to increase its efficiency. The new boilers were severely limited by their coupling to an undersized, secondhand engine. The *Richmond* was consequently underpowered and difficult to steer, especially in the confined waters of the James River.

Many criticized both the length of time required to finish the vessel and its secondhand machinery. Lieutenant John Taylor Wood, formerly stationed on CSS *Virginia*, wrote, "I doubt very much if she goes below the barrier [Confederate obstructions placed across the James River near Drewry's Bluff], for with her weak engine (80 horse) she would be mobbed and forced out of the channel. She will not steam more than 5 knots, if that."[24]

George Weber, a sailor on the Confederate navy school ship *Patrick Henry*, echoed those sentiments: "the worst thing . . . is that she carries bad engines."[25] Nevertheless, the *Richmond* became by far the strongest naval defender of the Confederate capital upon completion in November 1862. The other ironclads of the nascent James River Squadron (the *Fredericksburg* and the *Virginia II*) were not completed for another sixteen months.[26]

The *Richmond* was kept busy as the flagship until the completion of the other James River ironclads, but it saw only one actual battle toward the end of its career. Worry over the defective engine persisted, especially when considerations were afoot to send the ship on the offensive past the river obstructions placed below the city of Richmond. These were designed to prevent the approach of Union warships, but they also presented a formidable obstacle to any Confederate vessel trying to pass below them. A deep drafted and ponderous warship like the *Richmond* would especially have had difficulty. Army engineer Alfred Rives duly noted that "it would certainly seem of doubtful policy to send below an ironclad vessel with [an] inferior single-acting engine."[27] Perhaps there was an upper-level realization of this "doubtful policy," for the *Richmond* was never called on for such a risky venture while operating by itself.

Figure 4.3. The only known photograph of a *Richmond* class ironclad: CSS *Chicora* at its wharf in Charleston, South Carolina, circa 1863. The bow is facing to the right of the image. Note the hand-operated bilge pump on the after flush deck, just above the bow of the ship's boat. The white rectangular structure to the left is an outhouse. On the top deck, twin pilothouses are barely visible behind the two groups of men.

The *Richmond*'s commander for most of its career was a Virginian and a veteran of thirty-two years in the US Navy. Robert B. Pegram gained some prestige in Confederate service early on from his command of the commerce raider CSS *Nashville* (not to be confused with the later ironclad *Nashville* at Mobile). His next command was altogether different: not just the ship but the whole strategic situation of the James River Squadron. That bottled-up fleet was not able to make real use of his experience.[28] Pegram and his other officers recorded little to nothing concerning the daily routine on board the *Richmond*. Indeed, they had little to do for the better part of the one and a half years Pegram was in command. The several temporary captains after Pegram's transfer also recorded nothing about the technical aspects of their vessel.

A bit of information concerning typical conditions encountered in the machinery spaces of the *Richmond* has come down in the accounts of Chief Engineer Henry X. Wright. He was the senior engineer of the James River Squadron and, judging by his record, an exemplary officer. Like so many other Confederate engineers, Wright, born in 1835, was originally from Norfolk, Virginia. He was a third assistant engineer in the US Navy before joining the Virginia navy with the same rank in June 1861. After distinguishing himself in the Richmond area in various duties at Drewry's Bluff on CSS *Patrick Henry* and CSS *Richmond*, Wright was rapidly promoted to second, then first, assistant engineer. He was awarded the rank of chief engineer for gallantry and good conduct on June 10, 1864, and was stationed on the James River flagship CSS *Virginia II* until the war's end.[29]

Wright's reports reveal that the usual size of the engine room and fire room crew on the *Richmond* was ten men and that he held weekly inspections of that vessel's machinery along with all others in the squadron.[30] He also recorded recommendations and other interesting points. For instance, in a report to Flag Officer John K. Mitchell dated October 21, 1864, the chief engineer stated, "In regard to the engineers, I would state that she should have another assistant. . . . Besides, the engine is a very unhandy, hard-working one, and the engine room very hot and uncomfortable. I would state that the temperature has been as high as 150° Fahrenheit during last summer, and is considered to be one of the hottest engine rooms in the Confederacy."[31]

Mitchell was appointed to command the James River Squadron on May 6, 1864, after serving as head of the Office of Orders and Detail. Unlike his command at New Orleans in 1862, this position made him much better able to respond to problems with his vessels. The facilities of Richmond were close, the river was well protected by obstructions, and, most important, Mitchell had time to make sure his ironclads were completed and prepared for battle.[32] The flag officer and his superiors were determined not to let the *Richmond*, the *Fredericksburg*, and the *Virginia II* share the fate of the *Louisiana* and the *Mississippi*. Instead of waiting for the Union forces to come to them, the Confederate naval forces in this case planned an offensive. Concerns over the *Richmond*'s engine being able to function in battle were therefore well justified.

That "unhandy" single-cylinder engine in fact appears to have given remarkably little trouble despite the hellish conditions listed by Wright. It apparently functioned well even during the Battle of Trent's Reach on January 24, 1865, which was the only significant engagement fought by the James River ironclads. In that battle Mitchell finally led the squadron downriver to attack federal vessels (including the powerful double turret monitor USS *Onondaga*) and supply depots. Overcautious pilots, low water levels, and difficulty in passing through the river obstructions threw the effort into confusion. Both the *Richmond* and the flagship *Virginia II* grounded repeatedly because of their deep drafts and became exposed to Union shore batteries. Having lost the element of surprise and suffering the loss of two unarmored vessels, Mitchell was unwilling to further risk his valuable ironclads. After taking heavy damage they were finally refloated, and they steamed slowly back upstream, never to venture forth again.[33] Not long afterward, on April 3, 1865, they were all destroyed after the abandonment of the Confederate capital.

The *Richmond*'s remains in particular apparently still lie fairly intact under the mud of the James River. An archaeological survey may reveal more clues about the construction of this first ironclad of its class and the configuration of its half-new, half-scavenged power plant. The interesting career of the engine's source, USS *Arctic,* or Lightship No. 8, also warrants further investigation.

CSS *Chicora*

The first ironclad built for the defense of the strategically important Charleston Harbor and the second of the *Richmond* class was launched on August 23, 1862. CSS *Chicora*'s design is known to have followed the *Richmond*'s in most respects, since it was one of the very few Confederate ironclads to be photographed (fig. 4.3). By all accounts the *Chicora* was well constructed and was considered among the best-maintained ironclads in the Confederacy.[34] Unfortunately, its performance was hampered by a secondhand engine.

The *Chicora* was built by J. M. Eason & Brother, which also furnished the ironclad's two new boilers (of the same type as others in the class) and auxiliary machinery. The Eason foundry and machine shop was one of the oldest and largest industrial works in Charleston, South Carolina, by the start of the Civil War. It was founded sometime around 1819 by a partnership between Thomas Dotterer and someone named Connor. A few years later Robert Eason entered the partnership, and by 1847 his two sons, James M. and Thomas D., were in control of the firm.[35]

The foundry quickly grew into one of the most competent in Charleston and assembled the first steam locomotive in South Carolina, the October 1830 *Best Friend of Charleston,* as well as at least ten others.[36] The establishment was manufacturing sawmill engines and dredging Charleston Harbor with a specially built dredge boat by the 1850s. Under the direction of the Eason brothers, the foundry expanded to include cupola furnaces, forges, lathes, cranes, vises, drills, and saws. One special machine was a six-thousand-dollar lathe made to order in Scotland and claimed to be the most powerful in the South. It could turn and finish a flywheel up to 12 feet in diameter.[37] Although their collection of tools and experience was impressive, the Easons were still not able to manufacture a new engine for the *Chicora.* Therefore the trend of procuring secondhand propulsion machinery for an otherwise completely new vessel continued.

The *Chicora*'s engine was salvaged from the 147-ton tugboat *Aid,* launched in August 1852 at Kensington, Pennsylvania. This small vessel, also called the *Concord* (possibly a previous or yard name), was built by Theodore Bireay and measured 91 feet long, 22 feet wide, and 8 feet in draft. The *Aid*'s engine was a single-cylinder vertical-inverted direct-acting type measuring 30 inches in diameter and 28 inches in stroke. It was low-pressure condensing, keeping in accordance with most coastal and seagoing steam plants. The piston rod measured 3.5 inches in diameter, and the crank shaft was 7.25 inches in diameter. A single two-furnace return flue boiler, 21 feet long and 7.5 feet in diameter, and a single four-bladed screw 8 feet in diameter completed the machinery outfit. Materials from these latter components, especially the propeller and shafting, may have been reworked later by the *Chicora*'s builders for use in the new ironclad.[38]

After its launch the *Aid* was taken to Philadelphia for machinery installation. The builder of the machinery as well as all other ironwork in the boat was the ubiquitous and accomplished Reanie, Neafie & Company.[39] Thereafter the *Aid* remained out of the historical record (except for a systems overhaul and boiler replacement in May 1855). The tugboat was in Charleston at the outbreak of the Civil War and was seized by the Confederates. Its engine served the *Chicora* well when installed in the ironclad in late 1862, but as in other cases it was underpowered for the size of the ship. The *Chicora* could not generally make more than 7 knots.[40]

The *Chicora*'s engine actually seems to have given good service. The ironclad generally only steamed back and forth in Charleston Harbor on guard duty and saw little action. Duty on the Charleston station was not entirely uneventful, however.

The *Chicora*'s most noteworthy commander, at least from an engineering and technological point of view, was a man already well versed in ironclad warfare. Lieutenant Alexander Warley of CSS *Manassas* fame returned to his native South Carolina after his prisoner exchange from Fort Warren. He was appointed the captain of the *Chicora* in July 1862 after time as executive officer on its sister CSS *Palmetto State*. Warley kept busy in addition to the monotonous picket duties by supporting boat raids on federal targets, and he even provided volunteers for the submarine *H. L. Hunley*. Warley's only major action as the *Chicora*'s commander involved bombarding a Union force attacking Fort Sumter by boat (the Union forces were trapped against the walls of Sumter and forced to surrender). He did not have anything to say about his new command's performance, and by October 1862 he was detached for a staff position ashore. Thus Warley missed the *Chicora*'s one chance for glory in January 1863.[41]

Some of the *Chicora*'s other officers left valuable testimony pertaining to their ironclad. The career of one young engineer in particular reflected the broad range of new naval warfare technologies (ironclads, torpedoes, torpedo boats, and submarines) being developed by the Confederates. Fortunately, he left memoirs that have now been published.

James H. Tomb, born in Florida in 1839, became extensively involved with the new weapons in the Charleston area during his wartime service. By the time of his appointment to the *Chicora* as the third assistant engineer, Tomb had already participated in the momentous events of the Mississippi and New Orleans theaters on board the warships CSS *Jackson* and

CSS *McRae*. He was captured after the Battle of New Orleans and sent to Fort Warren in Boston Harbor before being paroled at an uncertain date. Tomb returned to Confederate service and was next ordered to Charleston for appointment to the city's first ironclad. He was on board the *Chicora* during that ironclad's most active months, but his greatest contributions to the war effort came after he was attached to the famous semisubmersible torpedo boat CSS *David* as an engineer and later as the commander. He was subsequently promoted to chief engineer for gallantry and finished his wartime career placing mines in the Savannah River.[42]

Regarding the *Chicora*, Tomb recorded his impressions of the machinery as follows: "The engine was single acting and was taken from a tug. While the engine was lacking in power, she was a good ship, after so many tinder boxes and cook shops we had had. She was thought to be able to make eight knots without forced draft, but when completed, could not under the most favorable circumstances make over seven."[43] These observations, like those pertaining to more unusual designs like CSS *Louisiana*, emphasize the difference between theory and real-life performance of ironclad steam engines. Hence even the fairly conventional hull designs of *Richmond* class ships like the *Chicora* suffered from miscalculations of suitable motive power generation.

Although Tomb's memoirs are fairly brief and often lack detail, he did provide an eyewitness account from the engine room during the *Chicora*'s and the *Palmetto State*'s only sortie from Charleston Harbor. Under cover of darkness on the night of January 30–31, 1863, the two ironclads moved out to attack the Union blockading fleet. Concerns over a lack of power prevented the *Chicora* from attempting to ram any target vessels. Tomb explained as follows:

> It was decided to make an attack on the blockading fleet off Charleston. January 30, 1863, we all had steam up, and about midnight proceeded over the bar. . . . We made about seven knots and I think the *Palmetto State* made about eight. About daylight on the 31st, we started for the blockading fleet. The intention was for both ships to ram, as it was felt we could hope to do nothing with them [enemy ships] if they had a chance to get away. . . . The *Chicora* . . . headed for a ship and we in the engine room were waiting for the signal to back full speed, . . . [but] we were told we would not ram. . . . They feared the *Chicora* would not have power to back clear again.[44]

The ship the *Chicora* engaged was the steamer USS *Keystone State*. The Confederate ironclad had difficulty closing in because of its slowness, but a shot through the Union ship's boilers scored the knockout blow. The *Keystone State*'s engine room crew was, however, able to affect quick repairs, and the ship escaped after an initial show of surrender, enraging the Confederates. The heavy *Chicora* was still unable to catch the badly damaged Union ship, and the next morning it returned empty-handed to Charleston with the *Palmetto State*.[45]

As a result of the Confederate attack, the blockading fleet was driven off, but no enemy ships were captured or sunk. Neither ironclad received noticeable damage. Although it was a tactical victory, the raising of the blockade was only temporary because the Union ships soon returned to station. The stalemate of Union blockaders outside the harbor and Confederate ironclads within continued.

After January 31, 1863, the *Chicora* never ventured from Charleston Harbor again. The weak engine, though still mostly reliable, even there got the ironclad into trouble occasionally. In particular, as part of its duties the *Chicora* acted as a picket near the harbor entrance. Strong tides are commonly encountered there, and the ship was forced to cast anchor in the face of a 2-mile-per-hour ebb on at least one occasion. Even with the engine at full power to aid in keeping position, the *Chicora* continued to drag.[46]

As the weeks and months went by, the ironclad continued to perform picket and general guard duties, all the while in an increasingly worse condition. By 1864 its power plant was unreliable and nearly worn out; it was reported on January 8, 1864, that the *Chicora* "wants a new boiler."[47] Nothing could really be done to remedy such a crucial problem—the Confederates by that point could simply not spare the time, men, or materials for a proper refit. The *Chicora* (as well as the *Palmetto State*) lingered on in such a manner until the evacuation of Charleston before General Sherman's forces in February 1865. Like its sister ironclad, it was destroyed to prevent capture when Union forces took the city.

Sixty-four years later, the US Army Corps of Engineers surveyed the remains of the scuttled Charleston Squadron ironclads in July and August 1929. The findings seem to corroborate the contemporary propulsion system details given for the *Chicora*. The wreck believed to be this ironclad has an engine measuring 30 inches in diameter with a 26-inch stroke, turning a three-bladed propeller 8 feet in diameter.[48] The slight difference in dimensions is probably the result of working underwater in a murky environment like Charleston Harbor. Future surveys of the scattered wreckage at the site may uncover more details.

CSS *Palmetto State*

CSS *Palmetto State* was Charleston's second ironclad and the third *Richmond* class vessel. Although laid down in January, it was not launched

Figure 4.4. The Confederate gunboat *Lady Davis*, formerly the tugboat *James Gray*. Except for its iron hull, this vessel was typical of most mid-nineteenth-century tugboats, several of which were important sources of machinery for Confederate ironclads.

until October 11, 1862 (the *Chicora* was laid down in April and launched in August).[49] The *Palmetto State* was initially financed with money from the women of South Carolina before the Confederate government assumed responsibility, whereas the money for the *Chicora* was provided from in-state sources throughout construction. It remains uncertain how much these different financial situations contributed to the construction times of the two ironclads.[50] Monetary issues aside, the ships were ultimately identical in most respects and spent their entire careers together.

The *Palmetto State* conformed closely to the dimensions and tonnage of the other ships in its class but was completed with an octagonal casemate and the pilothouse placed abaft the stack. The reason for these design changes is unclear.[51] The ironclad was built by James Marsh & Son at its Charleston shipyard, with the foundry of Cameron & Company building and supplying the *Palmetto State*'s new boilers and auxiliary machinery.[52] Presumably the boilers were two of the same standardized kind fitted to the other *Richmond* class vessels.

Cameron & Company, widely known as Phoenix Iron Works after a major rebuilding after a fire in 1850, was exceeded in size and output only by the Eason establishment. The former's origins actually dated back to a certain John Johnson's air furnace of 1802. By 1860 the Phoenix works consisted of a two-furnace foundry with carpenter, pattern, boiler-making, finishing, and machine shops fitted with all the latest equipment.[53] When Archibald Cameron acquired control of the company early in the Civil War, the ironworks could boast of having manufactured steam engines, drydock components, and contracted work with the Charleston and Savannah Railroad.[54]

As with the *Chicora*, a lack of resources, tools, and time prevented the *Palmetto State* from receiving new purpose-built engines. Machinery was again obtained from another small craft. This was the South Carolina navy gunboat CSS *Lady Davis* (fig. 4.4), a single-screw iron-hulled tugboat originally named *James Gray*, weighing 161 tons, launched on February 15, 1857, at Philadelphia. This boat, also identified as the *Tompkins* (possibly a previous or yard name), measured 87 feet long, 22 feet in beam, and 9 feet 5 inches in draft. The manufacturer of the hull and power plant was Reanie, Neafie & Company. The machinery configuration was unusually large for such a small tugboat: two single-cylinder horizontal direct-acting engines measuring 30 inches in diameter and 26 inches in stroke, with a piston rod diameter of 3.5 inches.[55]

Existing documentation of the *James Gray* also reveals that the tugboat was equipped with two force pumps 2.5 inches in diameter and 26 inches in stroke. Steam was supplied by a single-two-furnace firetube boiler 18 feet long and 10 feet in diameter. Power was delivered to the engines at 30 PSI, which turned a four-bladed screw 8 feet in diameter at 63 RPM.[56] It is likely that some of these materials were later adapted for use on the *Palmetto State* by Cameron & Company.

The completed *Palmetto State* was like the *Chicora*: reliable but underpowered. Its twin single-cylinder engines were certainly a more powerful system than that on the *Chicora*, but they were still too weak to speedily propel a heavy 174-foot-long ironclad. First Lieutenant William H. Parker expressed a contemporary evaluation: "Her engines always worked well, and under favorable circumstances she would go seven knots per hour, though her average speed was about six." He reported that the Charleston ironclads "were fine specimens of men-of-war and would have done credit to any navy. They were well officered and manned. . . . They were the cleanest iron-clads, I believe, that ever floated, and the men took great pride in keeping them so."[57]

Parker, who was born in 1826, was well qualified to make such statements. He was another relatively young officer with extensive experience on ironclads and other steam warships. Parker's career, by the time he served on the *Palmetto State*, encompassed service in the US Navy during the Mexican War and in the Confederate navy on shore batteries, in command of the gunboats CSS *Dixie*, then CSS *Beaufort* (notably alongside CSS *Virginia* at Hampton Roads), and finally the *Drewry* of the James River Squadron. After his service on the *Palmetto State*, Parker's excellent record was capped by his appointment as the commandant of the Confederate States Naval Academy school ship *Patrick Henry*.[58]

In contrast to Lieutenant Parker's achievements, the *Palmetto State*'s fortunes remained fixed to the city of Charleston. There were months of picket duty interspersed with the occasional boat raid and punctuated only by the sortie of January 30–31, 1863. In that battle, the *Palmetto State* was successful in ramming its target, USS *Mercedita*, while totally disabling that ship with a shot through the boilers. Although the *Mercedita* surrendered, the Confederates were unable to actually claim the Union ship without additional boats or men for a prize crew. The *Mercedita*'s captain and crew were therefore sworn not to serve against the Confederacy until regularly exchanged; their ship was then left alone to retreat as it could. This was another instance of chivalrous idealism carried over from earlier times; it occurred at a relatively late stage of the war, and the Union warship appears to have abided by the terms.[59]

After returning into Charleston Harbor, the next morning the *Palmetto State* resumed the tedious duties it was to perform for the rest of the war. As with the *Chicora*, the constant usage and lack of maintenance eventually took a toll: by 1864 the power plant badly needed at least a complete overhaul, if not total replacement. Neither happened. It was reported on September 7, 1864, that "the *Palmetto State*'s boilers are out of order. She goes very slowly."[60] The ship was finally destroyed along with the *Chicora* at the fall of Charleston on February 18, 1865. Any future investigation of the remaining wreckage in Charleston Harbor may reveal more information on the ironclad's design differences and the configuration of its adapted machinery.

CSS *North Carolina*

The first ironclad finished in North Carolina became the state's namesake. CSS *North Carolina* was begun in June 1862 at Wilmington and was built for the defense of that city, the Confederacy's most important blockade-runner port. It is surprising to learn, therefore, that a vessel intended for such a crucial station was destined to be the most decrepit *Richmond* class ironclad commissioned.[61]

The *North Carolina* followed the standard Porter design of the *Richmond*, the *Chicora*, and the *Palmetto State* but suffered throughout its career from poor construction and a weak engine. The ship was built across the Cape Fear River from Wilmington on Eagles Island at the shipyard of Beery & Brother and completed in June 1863.[62] The *North Carolina*'s layout was apparently quite similar to others of its class—archaeological surveys in 1993 and 1995 have contributed greatly to the scant historical details in this regard. Although little is known about why the ironclad was so badly constructed—the Beerys were government contractors and usually did competent work—much more detail is available concerning the machinery installation and its wretched performance.

Like the other *Richmond* class ironclads, the *North Carolina*'s power plant was a mixture of scavenged and new materials. It appears that at least the two boilers were new, of the standard type used in the *North Carolina*'s sister ships. Hart & Bailey Iron Works of Wilmington manufactured and installed the boilers as well as the shafting and auxiliary machinery. This large facility first opened under the name of Polley & Hart in 1840 as a manufacturer of copper stills for the Cape Fear area's turpentine industry.[63] The company was joined in 1857 by machinist John C. Bailey after expanding through the 1850s and branching into ironwork. By 1860 production had further shifted away from copper, and Hart & Bailey supplied Wilmington with much of its ironwork along with repairing machinery and building pump engines. The works included several machine, pattern, and copper shops along with a foundry.

Throughout the Civil War, Hart & Bailey was heavily involved with ship construction. Everything, from bolts and machinery parts to complete boilers, was produced.[64] This Wilmington establishment was one of the few works in the Confederacy capable of heavy forging and was accordingly able to rework components from the former Lake Erie tugboat *Uncle Ben* into what appeared to be a sufficient propulsion system for the *North Carolina*. As in so many other cases with engines taken from tugboats, however, the reality fell far short of the expectations.

The *North Carolina*'s extremely poor performance was in no small part a result of the undersized engine taken from the *Uncle Ben*. The tugboat was previously a local fixture of the Wilmington defense forces for some months as a small gunboat on the Cape Fear River. It arrived in the area by a twist of fate at the opening of hostilities: chartered in New York on April 9, 1861, for a diplomatic expedition to Charleston, the small tugboat put into Wilmington for repairs, unaware of the fateful events at Fort Sumter on April 12–13. The *Uncle Ben* was then captured by Confederate forces.[65]

The *Uncle Ben* was a long way from home when it was selected as *North Carolina*'s engine donor. The tugboat was built in 1858 by the firm of Bidwell and Banta of Buffalo, New York. This shipyard was established in 1808 by Nathan Bidwell but was burned by the British during the War of 1812. Bidwell quickly rebuilt it, turning out the first steamship on the upper Great Lakes in 1818. In 1836 he partnered with Jacob W. Banta, and the yard grew to include a marine railway and the first stationary drydock on the Great Lakes. By the 1850s the facility was one of the largest on the lakes, turning out luxury steamships, dozens of schooners, and other steamboats.[66] The *Uncle Ben* was one of these vessels, built to serve the yard and the city of Buffalo in general.

Few specifications of the *Uncle Ben*'s construction or dimensions have been found. The tugboat apparently had a tonnage of 155 and probably reflected the general standard design of other mid-nineteenth-century boats of similar purpose.[67] The engine was built locally by Shepard Iron Works of Buffalo, but little else is known other than that it was ultimately too weak to propel the *North Carolina* effectively.[68] It was probably a typical single-cylinder vertical-inverted setup, commonly equipped to tugboats like the *Uncle Ben*.

Once the *North Carolina* was completed, it turned out that it could not make more than 2 or 3 knots. The ironclad was taken downriver to a defensive position at the mouth of Cape Fear, where it was used as a floating battery near Smithville (now Southport). The *North Carolina* was prevented from ever going to sea by its weak propulsion and deep draft. It could only be towed back upriver occasionally in an attempt to kill excessive marine growth on the hull. For its entire career Wilmington's first ironclad served uneventfully as a floating defense.[69]

The *North Carolina*'s problems were soon compounded when on station it became an easy target for marine borers. The abundance of salt water near the river mouth, coupled with a lack of copper bottom sheathing to cover the ship's unseasoned wood, allowed shipworms easy access to its unprotected bottom. Daily life on board the ironclad soon turned into a battle not just against boredom and indiscipline but against an increasing number of leaks.

In contrast to the similarly employed CSS *Georgia* at Savannah, the *North Carolina* has a contemporary account of daily life on board. Shortly after its completion in June 1863, Third Assistant Engineer Charles Peek was assigned to the crew. He dutifully recorded his experiences on the poorly built ironclad. Peek's accounts, mostly letters to his sister, provide a fascinating glimpse of service on a particularly unpleasant craft. The young engineer did not have much to write about beyond the occasional incident, excursion, or personal encounter, but he did provide rare technical details about his ironclad's construction and machinery. This is all the more valuable because primary documents pertaining to the *North Carolina* and its sister CSS *Raleigh* are quite scarce (almost nonexistent for the latter).

Charles Smith Peek was only seventeen years old when he joined the *North Carolina*'s crew. He was a Virginia native who spent the war's first two years in an artillery unit. The extent of his mechanical experience is unknown, but it was enough for an appointment as the third assistant engineer in April 1863. Peek was sent to Wilmington, where he first served on the *North Carolina* and then, after the ironclad's loss, the cruiser CSS *Chickamauga*. He finished the war as part of the Semmes Naval Brigade, an infantry unit composed of unshipped sailors.[70]

Once settled on the *North Carolina*, Peek did not have long to wait before his new ship got into trouble. The ironclad was still in Wilmington but soon began its journey downriver to Smithville. The unsuitability of the *North Carolina*'s 12- to 13-foot draft for the often shallow and always treacherous Cape Fear soon became apparent. On July 1, 1863, Peek wrote the following:

> We even got aground on yesterday [*sic*] on account of the tide being so low and we drawing so much water. After we got aground I bank[ed] my fires and [Second Assistant Engineer Joseph] Freeman took the first watch. . . . We [later, after the tide came in] raised steam and got off. After travelling about two hours we anchored off Smithville, the name of a little town at the mouth of Cape Fear River. We lay all night with our fire bank[ed] and this morning we [illegible] fires and blew off steam. . . . When I came out of the engine room last night I was ringing [*sic*] wet with presperation [*sic*].[71]

Matters soon settled into a dull routine once the *North Carolina* was on station. Heavy drinking (sometimes even on duty) began to be a problem. At the end of July Peek testified, "The two 2nd asst. Engineers on this steamer are very fond of whiskey; they sat up last night until two o'clock drinking."[72] The issue only became worse as the summer dragged on. On August 12 he wrote, "We expect to go up the river today to rill the worms and barnicles [*sic*] that have accumulated on our bottom line since we have been in salt water. One of our Engineers was suspended the other day for getting drunk while on duty."[73] Maintaining discipline on board became difficult because several of the *North Carolina*'s crew members and officers continued to indulge heavily. Apparently next to nothing was done about it.

Very little happened on board over the winter and into the spring. Peek noted holidays, sickness, and the general lack of supplies or things to do on the *North Carolina*, but he did not mention significant details about matters in his department again until July 6, 1864, well after the ship had begun to sag and leak noticeably. As in cases elsewhere, the Confederates were aware of the problems but were unable to spare men or resources for a proper remedy. It did not help that there was no drydock at Wilmington for a complete hauling out.

According to Peek's July 6 letter, the *North Carolina* was again briefly moved up to Wilmington and again encountered trouble negotiating the river on its way back to station:

> We left Wilmington on the 2nd of July and anchored off Smithville for the night of the 3rd. We got a log in our propeller as we were crossing the shoals and had to anchor. When the tide fell we were about 5 ft. out of water, that is we draw about 13 ft., and there was not but 8 ft. on the shoals. I thought that the weight on the

> shield would crush in and I pack[ed] up my cloths [*sic*] ready for a start but the *old ship* stood it well. We were towed off next evening by two tugboats and are now safely at anchor off Smithville. . . . I am getting tired of this old ship. I want to get on some ship that can get along without [being] towed everywhere she goes.[74]

Two days later Peek wrote, "We have to keep up steam to pump the ship out, it is very hard; we have to keep watch and it also makes the ship very warm."[75] The situation continued like this for another two months.

Late July 1864 was the beginning of the end for Peek's "old ship." His use of this descriptor easily expresses the "in reality still technically new" ship's dilapidated state. Peek's letter of July 24 described what was to be the *North Carolina*'s final situation and resting place: "Our old ship has nearly played out. We have her moured [*sic*] head and stern in shoal water that is at low water. She is about 12 inches from the bottom, and she leaks so badly that we have to keep steam up all the time. There is no discipline on her now."[76] A little over a month later, in early September 1864, the *North Carolina* sank at its moorings off Smithville.[77] The hull was so riddled by that point that the pumps simply could not keep up.

As the ship settled into the soft bottom sand, it lay partly awash with the casemate exposed above the surface. Access was easily gained for some time after the crew actually abandoned ship. Everything valuable was eventually stripped off, and the crew members were reassigned. The *North Carolina*'s spectacularly unsuccessful career was over.

The wreck's exposed location meant that it was easily picked apart over the years. Most of the valuable iron armor was removed in 1864 and 1868; the ship's machinery was also salvaged at an unknown time. In 1871 the majority of the remains were destroyed by a fire of uncertain origin.[78] Only the very bottom of the *North Carolina*'s hull is left as an archaeological site. Much information has been gleaned from the timbers, but the machinery components are nearly nonexistent. Debris and small artifacts associated with the boilers reveal their approximate location—the same as that shown in plans of sister ship CSS *Savannah*. No evidence of the engine mounts survives. Unfortunately, the remaining hull and scattered metal pieces will probably never reveal exactly why the *North Carolina* was such a crank ship compared to all the others of its class and why the *Uncle Ben*'s engine was such a complete failure.

The *North Carolina* is the last Confederate ironclad known to be provided with machinery from one of the great shipyards and engine works of the antebellum Northern states. This case perhaps emphasizes best of all the unsatisfactory nature of scavenged tugboat power plants, no matter how well constructed they originally may have been.

CSS *Raleigh*

CSS *Raleigh* was the second ironclad built in Wilmington but was not constructed in the same yard as the *North Carolina*. These two ships could not have been more different despite being of the same class and built in the same city: the *North Carolina* was poorly built and a slow failure, whereas the *Raleigh* was stated to be one of the finest examples of Confederate ironclad construction during its all-too-brief career. The ship was built at the yard of J. L. Cassidy & Sons at the foot of Church Street in Wilmington. Like the *North Carolina*, it closely followed the design and layout of the other vessels of its class, but apparently it was much better built and had a lighter draft. In addition, greater trouble was taken to find reliable motive power for the *Raleigh*. Archaeological investigations of the ironclad's wreck in 1993–1994 ultimately found that the engines were the single-cylinder horizontal direct-acting type called for in the original plans.[79] Their installation closely matches that shown in the plans of the *Raleigh*'s sister ship, the *Savannah*, which was constructed around the same time.

The exact source of the steam engines remains unclear. Initial attempts to procure a power plant from the wrecked blockade-runner *Modern Greece* failed, so new engines of the preferred horizontal type were sought. Those put in the *Raleigh* appear to have been built in Richmond, but whether they were built specifically for the ironclad is still open for debate. Questions remain because the 1993–1994 investigations of the wreck failed to find a builder's plate stamped onto or affixed to the machinery.[80]

The *Raleigh*'s power plant may have been manufactured in Richmond, since Tredegar Iron Works produced the armor and possibly the fittings for several North Carolina ironclads.[81] A direct reference to the *Raleigh*'s engines as newly manufactured was made by a contraband (an escaped slave) named William Robins, who apparently was involved with construction.[82]

There exists another significant possibility: that the *Raleigh*'s engines came from an unfinished wooden gunboat built at Washington, North Carolina. This boat and two others were under construction in 1861 and were approximately 150 feet long with a 25-foot beam.[83] All were destroyed before completion when Union forces began to move into coastal North Carolina.

The Washington gunboats and others like them required engines of similar size to the *Raleigh*'s. Although there is no direct evidence that these engines ended up on the ironclad, there is other circumstantial data. It is known that the engines for the 150 foot gunboats built (none were ever completed) in North Carolina and Florida were to be provided by Talbott & Brother of Richmond, more commonly known as the Shockoe Foundry; the Florida boats' engines were 26 inches in cylinder diameter with a 24-inch stroke.[84] These measurements only roughly approximate those found on the *Raleigh*. If the ironclad's engines came from a Washington gunboat, it is safe to assume that they were altered during construction. In contract work, the Confederate navy seems to have given builders considerable leeway in interpreting specifications.[85] Nothing can be stated with any certainty to date.

The source of the *Raleigh*'s boilers and auxiliary equipment seems clearer. The boilers were two, again of the standard low-pressure horizontal firetube type, built by the local Clarendon Iron Works. Thomas E. Roberts established the works in Wilmington early in the war and manufactured many iron components for ship construction. Roberts's firm was accordingly involved in naval construction in Wilmington, although it did not have the capability for heavy forging. Clarendon Iron Works also cast many boiler parts, including boiler iron, grate bars, flues, and fronts, suggesting that it built the boilers for the *Raleigh*.[86] In addition to the construction of boilers, the building of marine steam engines was another task entirely for small foundries like Clarendon. The construction of engines often had to be outsourced. Most of the current confusion over machinery sources arises from this wartime necessity.

For unknown reasons the *Raleigh* was not finished for more than a year after the *North Carolina*, even though they were both laid down at about the same time, in the spring of 1862. This may have been partly because of the aforementioned efforts to find better machinery. The vessel's motive power when finished certainly appears to have been above average. In any case, by March 1864 the *Raleigh* was nearing completion, and by April it was commissioned.[87] Immediately afterward the brand-new ironclad was called upon to make a sortie into open water against the Union blockade.

The *Raleigh*'s appointed captain had previously served on three other ironclads. First Lieutenant John Pembroke Jones (born in 1825 near Hampton, Virginia) was on CSS *Virginia* in 1862, then commanded CSS *Georgia* and CSS *Savannah* in 1863. He was next transferred to the James River Squadron in command of the gunboat CSS *Nansemond* before moving to the newly completed *Raleigh*. Jones also ended up commanding the *North Carolina* after that ironclad's aging and sickly captain, Commander William T. Muse (born in 1811), died in April 1864.[88]

Unlike Muse, Jones had plenty to do with his new ironclad, although he had it only a few weeks. He knew all too well how terrible it was being in charge of a ship (i.e., the *Georgia*) incapable of going anywhere. The *Raleigh* did not have that problem, and Jones soon had a chance to test the ship. On the night of May 6–7, 1864, the *Raleigh* (with the Wilmington Station commander, Flag Officer William F. Lynch, on board) steamed downriver as the escort of a blockade-runner and in the company of the wooden gunboats CSS *Yadkin* and CSS *Equator.*[89]

The actual objectives for the offensive remain unknown and were obscure even at the time of the incident. Lieutenant William H. Parker, Confederate States Navy, wrote, "What [she] . . . went out for has never been ascertained." Regardless of what its purpose might have been, on the evening of May 6 the *Raleigh* was in a position to surprise the blockaders. Beginning around eight o'clock the *Raleigh* attacked USS *Britannia* but was unable to keep track of that ship in the dark. In a running battle that lasted until about midnight, the Confederates intermittently engaged both the *Britannia* and USS *Nansemond* at long range. No hits were scored and both federal ships easily escaped the slow-moving ironclad, leaving it alone for several hours.[90]

Early in the morning of May 7 another battle began that lasted until approximately seven o'clock. The *Raleigh* simultaneously engaged the approaching blockaders USS *Howquah*, USS *Nansemond*, USS *Kansas*, and USS *Mount Vernon*. In another running fight at long range, the ironclad suffered no damage while superficially injuring the federal ships. Perhaps realizing that they were outnumbered in an exposed position, Jones and Lynch retired from the battle and the *Raleigh* retreated toward the Cape Fear River entrance at New Inlet. Although they may have simply been unwilling to risk the having the ship remain at sea between tides, the captain's and flag officer's decision to move to safe waters ironically proved disastrous: as the *Raleigh* began to turn upriver for Wilmington, it struck a bar known as the Rip and grounded fast.[91] Attempts to float the ship off proved futile, and as the tide ran out the weight of the *Raleigh*'s armor began to press down on the unsupported hull until it broke its back and sank. The pride of the Wilmington defense force had lasted less than one month in service.

The *Raleigh* was not immediately or extensively salvaged because of the dangerous conditions in the middle of the Cape Fear River. The wreck subsequently broke up, and little was saved from it, but the crew was not endangered because the weather was calm.[92] Later salvage efforts concentrated on removing what armor was accessible, as well as at least two of the guns. The boilers were also successfully removed and were sent to Columbus, Georgia, for fitting on the gunboat CSS *Chattahoochee*. The wreck site, soon rendered inaccessible by the shifting sands of the river, remained a navigational hazard during the remainder of the Civil War and for several years later. At least two incoming blockade-runners are known to have hit the submerged remains, and only by the closing of New Inlet in 1881 was the *Raleigh* allowed to rest in peace.[93]

After New Inlet was closed, the wreck was visited only by the occa-

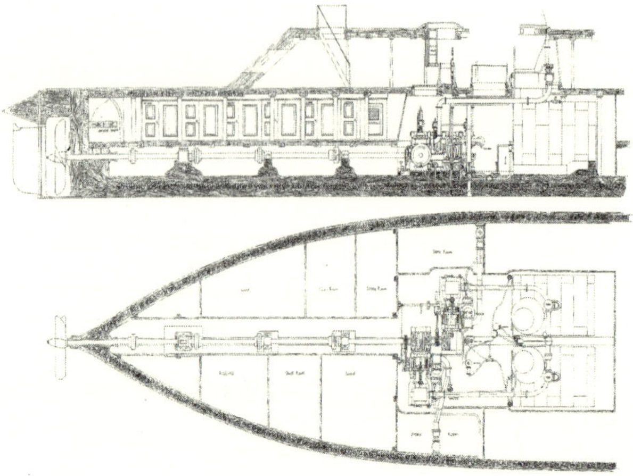

Figure 4.5. Machinery plans of the *Savannah*, redrawn by Bob Holcombe. Note the location of the ship's wheel and pilot hatch directly over the engine room, just behind the smokestack in the upper view. The two horizontal firetube boilers are forward, each with a main steam line going to each engine. The staggered arrangement of the horizontal engines within the hull and their connections to the central propeller shaft is demonstrated in the bottom view. This setup was used on virtually all ironclads with horizontal engines. Many other details—such as the propeller shaft couplings, the overboard discharge pipes, and even the layout of the officers' wardroom above the shaft alley—are also shown. These plans are perhaps the most detailed of any Confederate ironclad.

sional fisherman, who frequently lost nets to it. Beginning in the 1970s, several locals began to suspect that the site was the remains of the ironclad, but full-scale investigations did not commence until 1993–1994. Because of difficult conditions caused by tides, currents, and poor visibility, only the starboard side could be studied in detail, but the wreck appeared to be in surprisingly good condition.[94]

The investigations revealed new data on the *Raleigh*'s engines, which were previously thought to have been salvaged. They measure 24 inches in diameter with a 3-foot stroke. Each cylinder has a 3-foot-square built-in valve chamber and rests on concrete blocks instead of the more usual wooden ones. The piston rod diameter is 3 inches, the connecting rod diameter is 4 inches, and the propeller shaft diameter is 9 inches.[95]

The *Raleigh* remains relatively undisturbed to this day, mostly because of the dangerous conditions surrounding it that prevent easy access. It is probably the most intact extant example of a *Richmond* class ironclad. For this reason, coupled with a lack of firsthand information in the historical record, the *Raleigh* warrants further investigation. A top priority should be determining more details of its construction, such as any possible hull design changes resulting in the lighter draft and better seakeeping abilities claimed for the ship in contemporary accounts. A focused search for the engine builder's plate could also be revealing.

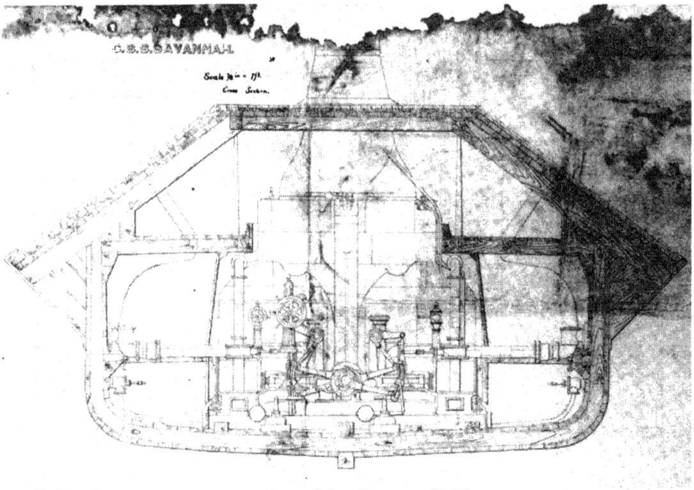

Figure 4.6. Engine room cross-section of the *Savannah*. The connecting rods, eccentrics, and links for the reversing gear can be seen on the propeller shaft at the center. On either side of the linkages are the engine cylinders. Note the engineer's platform with the throttle wheel and other controls at berth deck level on the left. The paired boilers and their uptakes to the smokestack are shown in the background. The ship's nearly flat bottom, shallow keel, and armored knuckle backed with solid timber at the waterline are all distinguishing characteristics of the *Richmond* class and many other Confederate ironclads.

Much remains to be examined, and answers to these questions may well be found some day.

CSS *Savannah*

The last *Richmond* class ironclad completed during the Civil War was constructed in Savannah, Georgia, and named for that city. CSS *Savannah* was built at the shipyard of Henry F. Willink and closely followed all its sister ships in design and layout. The *Savannah* was probably the best constructed of any ship in its class, and it had new boilers and engines. By all accounts this power plant was reliable.

Much information exists concerning the *Savannah*'s engines, boilers, and hull construction, because both the ship's and the engineering plans are preserved in the collections of the Port Columbus Civil War Naval Museum. The plans offer a comprehensive look at the ironclad's machinery installation (figs. 4.5 and 4.6).

Records indicate that the *Savannah*'s engines were originally intended for a 150-foot gunboat built by Frederick G. Howard at Milton, Florida, but when that project was cancelled Secretary Stephen Mallory ordered them transferred to the ironclad then being built in Georgia. In a letter to Chief Engineer James H. Warner on April 10, 1862, the secretary wrote, "You will take charge of this work at once, and are authorized to make such alterations and additions to the engines as will adapt it to one of the iron clad gun boats to be constructed by Mr. Willink [the *Savannah*'s builder]."[96] The vessel to which Mallory referred could only be the *Savannah*, since the other ironclad built by Willink, CSS *Milledgeville*, was not laid down until much later.

Measurements taken from the *Savannah*'s engineering drawings show that its two single-cylinder horizontal direct-acting engines were 28 inches in bore diameter (34.5 inches in external cylinder diameter) with a 20-inch stroke. Unlike the purpose-built engines of CSS *Raleigh*, which were also probably obtained from an unfinished wooden gunboat, the *Savannah*'s valve chambers were on top of the cylinders. It appears that reverse motion was accomplished by the tried-and-true Stephenson's valve gear. The propeller shaft was 50.3 feet long, 8 inches in diameter, and manufactured in five sections. The *Savannah*'s two boilers were of the preferred low-pressure horizontal firetube type. They were 7 feet 10 inches in diameter, 11 feet 2 inches long, and 12 feet high. All components were constructed by the Columbus Naval Iron Works.

The facilities for engine and shipbuilding in Columbus, Georgia, came to represent the second greatest concentration of industry in the Confederacy. Skillfully managed and well furnished with the necessary equipment, Columbus Naval Iron Works manufactured complete sets of machinery for at least five ironclads (the *Savannah*, the *Columbia*, the *Jackson*, the *Milledgeville*, and the *Wilmington*) and also refurbished old machinery for installation in new vessels. The Columbus works became the single most important Confederate source for purpose-built engines during the war and was exceeded in industrial output only by Tredegar.

Columbus was located at the head of navigation on the Chattahoochee River and was ideally suited for industrial development. Cotton and other agricultural products were shipped along the rivers, and a thriving steamboat-building industry combined with cotton and wool mills turned the city into the most industrialized in the Southern states after Richmond.[97] It was there that a gentleman named William R. Brown established a small foundry, Columbus Iron Works, on the lower town riverbank in 1848.[98] This small establishment constructed steamboat engines, boilers, and fittings using only one rolling mill and forge. Brown's works remained at best a minor operation, despite his accomplishments, until 1861. The coming of war changed everything.

By July 1861 Brown had begun turning out small cannons and accepted contracts from the Confederate Office of Ordnance and Hydrography. In January 1862 the Columbus works successfully produced several types of weapons, including mortars and rifled cannons. Noting the foundry's ideal site on the Chattahoochee riverbank, the Navy Department leased Brown's facility beginning in September 1862.[99] Now known as the Columbus Naval Iron Works, the firm grew into a large naval industrial complex sending marine machinery, naval ordnance, and engineering expertise throughout the Confederacy.[100] The man tasked with managing this operation was Chief Engineer James Warner, probably the Confederacy's most skilled engineer. Warner was put in charge of the machine shops in Columbus and oversaw the manufacture and refurbishing of a large number of steam engines. He collaborated extensively with Lieutenant Augustus McLaughlin, Confederate States Navy, who was an able administrator and the superintendent of the adjacent Columbus Navy Yard, where a few of the recipients of Warner's engines were built, including the ironclad *Jackson*.

Both McLaughlin and Warner were remarkable men, and their skill and unwavering work ethic led to the exponential growth of Columbus Naval Iron Works during the war. They were close in age (McLaughlin was born around 1823, and Warner in 1827), and both were originally from a border or Northern state: Maryland and Ohio, respectively. McLaughlin had served in the US Navy since 1846, reaching the rank of lieutenant shortly before his resignation to join the Confederacy in April 1861. He initially received the same rank in the Virginia and Confederate navies but was soon promoted to first lieutenant. McLaughlin spent the war mostly in administrative positions: on board the old 1797 frigate and receiving ship CSS *Confederate States* (formerly USS *United States*) at Norfolk; at Apalachicola, Florida; and for the rest of the war commanding the Columbus Naval Station.[101]

Warner had achieved the rank of chief engineer in the US Navy by the time he resigned his commission. His Confederate sympathies probably arose in part from a long residence in Portsmouth, Virginia. Warner's highly valuable skills meant that he was called to serve at several different theaters during the war: first at the Gosport Navy Yard and in New Orleans in 1861, at Savannah in 1862, and finally at Columbus, where he spent the remainder of his career. It was there that he made the greatest contributions to Confederate steam engineering. He designed and drew plans for new engines, rebuilt and modified old and secondhand machinery, and gathered suitable materials and components from all over the Southern states. At the war's end, Warner was a great help to the occupying federal forces in listing and gathering up all those parts formerly intended for Confederate warships. He was shot and killed for unknown reasons by a Union soldier in February 1866.[102]

Warner's leadership was a huge boon to the naval war effort. By 1865 the Columbus works had grown to become one of the largest in the Confederacy. The armory contained a small rolling mill, a nearly complete large rolling mill, and one large steam hammer from Mobile.

The machine shop contained three small and two large planers, sixteen iron lathes, one large lathe, three drill presses, and thirty vises with anvils.[103] Also present on the site were a blacksmith shop with ten forges, a foundry, a boiler shop, a copper shop, and a pattern shop. Entire sets of machinery were turned out along with a number of rebuilt engines. All of this was run by a force of more than three hundred men, with auxiliary facilities in Prattville and Eufaula, Alabama. Personnel from Columbus were sent all over the Confederacy and maintained offices in Charleston, Savannah, Selma, and Mobile, among others. These local outlets were often in charge of assembling and installing the batches of ironclad machinery sent from Columbus. For example, Alvin Miller's foundry in Savannah, which built CSS *Georgia*, was in charge of assembling and installing the machinery for the Savannah ironclads, including the unfinished *Milledgeville*.[104]

Columbus was not captured until April 16–17, 1865, more than a week after Appomattox. The Confederacy's second largest industrial site continued production up until the very end. It successfully gathered and contributed a considerable amount of expertise in marine engineering during that time. In the end, Columbus Naval Iron Works (and its satellites in Prattville and Eufaula) succeeded in providing power to a number of ironclads and wooden gunboats as well as manufacturing naval ordnance. In doing so it had a huge positive effect on the Confederate war effort.[105]

Perhaps Columbus's first successful product was the *Savannah*'s power plant. The ironclad was launched on February 4, 1863, and served as the flagship of the local naval forces.[106] A few different in-service accounts of its machinery exist. Some of the most important firsthand details come from official Savannah Squadron correspondence papers now housed at the Manuscript, Archives, and Rare Book Library at Emory University in Atlanta. All the documents generally praise the ironclad's performance, although during trials in July–August 1863 some troubles were encountered.

Much data was gathered and reported during the *Savannah*'s trials. They were initiated when Secretary Mallory wrote on July 18 to Flag Officer William W. Hunter, the commander of the Savannah Squadron: "You are requested to have the 'Savannah' occasionally run up and down the river, not only to perfect a test of the machinery, but to exercise the Engineers and Firemen and perfect an organization; and you will direct a report to be made thereon."[107] Hunter (born in 1803) was another aging captain appointed to command a Confederate naval station. His previous wartime duties included time as the master of the gunboat CSS *Jackson* at New Orleans and the commander of the naval defenses of Texas. Unlike several of his contemporaries, Hunter appears to have been a competent leader and administrator.[108] It is fortunate that many of his office's correspondence records still survive.

Apparently the same day that Mallory wrote Hunter's orders for the *Savannah*, the new ironclad was already undergoing trials. A report was given listing some of the ship's design problems. It is particularly interesting because it relates to troubles with one of the vessel's auxiliary steam systems: "The pipe from the steam [bilge] pump does not lead sufficiently far aft—as when it 'sucks' it leaves 6½ inches water in the ship, while with her flat floor [hull bottom] endangers the stores aft. The pipe should be extended aft to the pump well and then it will free the ship."[109] Presumably efforts were soon made to correct the pump defects, for they are not mentioned again. Before long, other more unofficial accounts were made of the *Savannah*'s first run. Crew member Henry Graves wrote to his sister Cora, "Today we took the 'Savannah' down river to test her engines. She started off beautifully and glided down stream as smoothly as if she were some water fowl. It was a strange sight to see that huge mass of iron, shooting from the water like the roof of a house, steaming along so like a thing of life."[110] Although this description may be unsuitably elegant for such an ungraceful ship as an ironclad, it does reveal that the *Savannah*'s steaming qualities were unusually good.

Unfortunately, in some aspects the machinery provided for the *Savannah* still reflected some of the early quality-control problems endemic to the steam engineering business, despite the skill of Columbus Naval Iron Works. Troubles were soon encountered and proved to be serious. On another trial run, one of the *Savannah*'s cylinders was broken by a rivet left in the steam passage.[111] Captain Robert F. Pinkney wrote to Flag Officer Hunter on August 3, 1863, "While this vessel was returning from the [torpedo] barrier, whither I had run her, in obedience to your order, to try her machinery, the starboard engine became disabled from the breaking of the piston. By subsequent investigation, it has been discovered that the steam-cylinder is also broken."[112] In other words, the machinery breakdown had crippled the ship.

Hunter wrote an order the next day for a thorough investigation of the broken engine by the senior engineers of the Savannah Squadron. Recommendations on repairs were also requested. Acting Chief Engineer John M. Tynan, another Virginia native and a Portsmouth resident who earlier in the war served on CSS *Virginia* (notably during the Battle of Hampton Roads) and CSS *Chattahoochee*, was in charge of the investigation.[113] He made his report to Flag Officer Hunter at the end of the day on August 4:

Sir: In accordance with your order herewith appended, we have held a strict and careful survey on the engines of the C.S. Steamer Savannah and most respectfully submit the following report of their condition for service, of the damage done to the starboard engine, etc. We are of the opinion that the ship can be worked by the port engine as it is, by careful management. But as there will be some difficulty in starting the engine, in consequence of the liability to stop in the centers [common to short-stroke horizontal engines], we consider it necessary to have a steamer in attendance to render any assistance that may be required; and for further security, we respectfully recommend that a trial be made at the wharf which will determine whether by some slight alteration of the [main admission and exhaust] valve, the working of the engine may not be considerably improved. Other than this we would not advise the engine to be used except in case of great emergency. The damage done to the starboard engine is as follows, viz: the cylinder, piston head, follower and rings are irreparably broken, outboard cylinder head cracked and piston rod bent. From the appearance of the broken parts of the piston, we are satisfied that—with the exception of about two square inches—it was broken entirely through the middle before being used on this vessel, and are of the opinion that it was done in keying it on the rod before it left the shop. We were shown a rivet 5/8-inch diameter and about 2½ inches long with a head of about 1-inch diameter which according to the statement of the engineers in charge was taken from the inside of the cylinder amidst the fragments of the piston. This rivet in our opinion was the immediate cause of the accident; we suppose it to have been left through the carelessness of some of the workmen in the steam passage of the cylinder, and by the rush of steam or some jar was carried down into the cylinder, and as there was only from ¼ to ⅜ of an inch clearance between the piston and cylinder head a breakdown was inevitable. Had the piston been perfectly good before steaming, the presence of this rivet in the cylinder would in all probability have caused the same amount of injury to the engine. The breaking of the cylinder was evidently caused by the parts of the piston broken off being acted on by that part still attached to the rod as the engines according to the statement of the "engineer of the watch" made two or three revolutions after the occurrence of the accident. The following repairs are necessary to put the machinery in good working order. Viz: A new cylinder, new piston with follower, rings, etc., a new cylinder head and the piston rod to be straightened. We think the most expeditious manner of repairing the engine would be by having it

> done at the "Naval Iron Works" Columbus, GA, where the engines were built. The time necessary to make the repairs will in our judgment be about one month.[114]

This report represents one of the most detailed accounts of the functioning of a typical ironclad engine and also reveals the kind of quality-control problems encountered by the Confederacy's nascent marine engine industry. It was quickly agreed to by both Flag Officer Hunter and Secretary Mallory, and the broken engine parts were sent to Columbus. The whole incident must have caused engineer Warner considerable embarrassment when he learned the reason one of his products was sent back.

The repairs went well, and the *Savannah* was successfully repowered within two weeks.[115] There are no further reports on the ship's performance under steam. The *Savannah*'s career mirrored that of so many other Confederate ironclads: steaming back and forth along the waterways it was built to defend, deterring any federal attack by sea. Captain Pinkney (born in 1812 in Maryland), a veteran of the 1862 debacle at New Orleans, could at least be satisfied that his ship was well constructed and capable of moving under its own power, unlike the *Savannah*'s squadron mate, the *Georgia*.[116]

By mid-1864 the land war began to progress toward the city of Savannah. Union General William Tecumseh Sherman began his devastating march across Georgia to the sea. Resources such as coal for the navy began to be cut off. Among the last accounts relating to the *Savannah*'s machinery are entries in the journal of sailor Robert Watson, who served on the ironclad after his transfer from the army. On May 1, 1864, he revealed the severe shortage of coal fuel, although by that time it was casually accepted: "Took some wood and got up steam and dropped out into the stream."[117] The *Savannah* and many other mid-nineteenth-century steamers could burn either coal or wood; the grate bars in the boiler furnaces simply had to be switched for the appropriate fuel material.

At the end of December Sherman's forces reached the city of Savannah and soon captured it. The city's namesake ironclad was left to cover the evacuation of Confederate forces on the night of December 20–21. The other Savannah Squadron vessels either fled upriver (small gunboats with light draft) or were destroyed to prevent their capture. Only the ironclads *Georgia* and *Savannah* remained. The former, unable to move, was soon bypassed and scuttled. The *Savannah* alone continued to bombard Union artillery units at long range all day on December 21. Watson reported on the danger of shells entering a weak point, the ironclad's smokestack: "The Yankees opened fire on us from the city. We were not slow in returning the compliment but with what effect I cannot say. The Yankees made excellent shots, nearly every one struck our sides or smoke stack. One shell went down the smoke stack and rested on the grating but did not explode."[118]

Once the gunnery duel ended and the Confederate retreat was known to have succeeded, the *Savannah* followed in the path of so many other ironclads, and on the night of December 21, 1864, it was fired by its own crew. The last *Richmond* class ironclad sank in a deafening explosion that shattered windows several miles away.

What remained of the *Savannah* was heavily salvaged in the 1890s. In those operations the ship's boilers and machinery were removed.[119] For this reason alone it is very fortunate that the *Savannah*'s detailed machinery plans survive. They provide invaluable information not just on this ironclad's systems but also on those of the other members of the *Richmond* class and of Confederate ironclads generally. In contrast, it would seem there is little left at the wreck site. Future archaeological investigations are necessary to determine what exactly remains.

Conclusion

The six *Richmond* class ironclads showed that the Confederate navy was making progress both in ironclad design and in obtaining adequate machinery. Although these vessels did have an unsuitably deep draft for most stations, with one exception they performed their intended duties well. The later examples in particular (the *Raleigh* and the *Savannah*) were considered excellent warships. All benefited from a strong design and unity of purpose: harbor defense. The *Raleigh* and the *Savannah* also benefited from new machinery, which was not originally intended for them but was still suitable. The earlier examples of the class unfortunately had to make do with secondhand power plants taken from small screw steamers. Although the *Richmond*, the *Chicora*, and the *Palmetto State* functioned about as well as could be expected, the weak machinery on the *North Carolina* was a major contributor to that ironclad's total failure. This underscored the general unsuitability of scavenged engines. Ironclads needed something better; they were bigger and much heavier than tugboats, lightships, and other craft.

The first type of standardized ironclad constructed for Confederate service reaped some early benefits of the slowly maturing marine steam engine industry. Facilities such as Columbus Naval Iron Works in Georgia especially made great contributions. Because of skilled leadership and engineers, the Columbus works was ultimately successful in providing ironclads in several Southern ports with refurbished or new machinery sets. The CSS *Savannah* was among the first. Confederate steam engineers on both land and sea were quickly learning, adapting, and improvising to meet the demands of war.

The operational experiences of the *Richmond*, the *Chicora*, the *Palmetto State*, the *North Carolina*, the *Raleigh*, and the *Savannah* paved the way for even better designs. A settled policy of primarily harbor defense meant that Constructor John Porter and his peers could concentrate on refining the traits desirable for their ironclads. Several characteristics of the *Richmond* class hull design and general construction were subsequently used in the next several ironclad so-called classes (none were actually formally grouped) built by the Confederates. These included the most formidable warships constructed in the Southern states.

FIVE

THE *TENNESSEE* CLASS IRONCLADS

Quite equal to the best of the kind built by the rebels.
—Rear Admiral John A. Dahlgren, US Navy

The large *Tennessee* class vessels were the next standard hull ironclads to be designed. They contained several significant improvements over the previous *Richmond* class, based on those ironclads' construction issues and operational experiences. Longer fore and aft flush decks for equipment stowage and working room combined with a slightly improved draft for their size ensured better handling, and heavier armor and armament made the *Tennessee* and the *Columbia* truly formidable warships. A third member of the class, the *Texas*, was laid down in Richmond but never completed.

The *Tennessee* and especially the *Columbia* were considered fine warships. Although technically sisters, or nearly sisters, their construction was spread out over the last three years of the Civil War, and the finished products differed greatly in configuration. The basic hull form stayed the same, but the *Columbia* in particular incorporated all the best features of late war ironclads, including new purpose-built machinery. Therefore the finished *Tennessee* class ships exemplified the evolving trends in Confederate ironclad design and changes in operational and material influences.[1]

The roles that the *Tennessee* and the *Columbia* played in the naval war were also quite different. Whereas the former achieved fame in one of the Civil War's largest naval battles, the latter ultimately made very little contribution to the Confederate cause. Fortunately, significant details exist describing the specifications and careers of these two ships. By all accounts they were among the first large ironclads provided with adequate machinery.

CSS *Tennessee*

CSS *Tennessee*, constructed in Selma, Alabama, became one of the Confederacy's most famous vessels. This ironclad was designed to be driven by a single screw, but serviceable engines were lacking. The otherwise

Figure 5.1. This port quarter view of the *Tennessee* off New Orleans was probably taken in 1865 after its capture and induction into the US Navy. The ship's unique design features are visible despite the abundance of deck awnings and other clutter. Note especially the long fore and aft flush decks, the four broadside gunports, the tall smokestack, and the large pilothouse integral with the casemate.

completely new ironclad was fitted with secondhand boilers and engines. In an interesting example of makeshift engineering, riverboat engines were installed and geared to the propeller shaft, which actually worked comparatively well in service.[2] This was a measure of necessity despite the obvious disadvantages of adapting high-pressure riverboat machinery (designed for paddlewheels) to turn screws. Boilers and engines from river steamboats were put into several Confederate ironclads simply because they were readily available. It was intended that these ships, including the *Tennessee*, be fitted with better purpose-built engines once they became available, but this optimistic plan was never fulfilled.[3]

The *Tennessee* was built by local shipbuilder Henry D. Bassett at Selma and was taken to the strategic port of Mobile, Alabama, for completion. The powerful ship was launched on February 28, 1863, but was not finished for another year because of shortages of men and materials and modifications made during construction.[4] Once completed, the *Tennessee* differed from Constructor John Porter's original design by having a shorter octagonal casemate with six guns. Its dimensions were 217 feet in overall length, 209 feet on deck, 48 feet in beam, and an average draft of 14 feet.[5] Fortunately, several photographs exist of the *Tennessee* after its capture by the US Navy in August 1864, revealing the overall appearance of the vessel (fig. 5.1).

The exact origin of the *Tennessee*'s machinery is uncertain, but there are enough historical clues for some conclusions to be drawn. Often cited as the source of the engines and possibly the boilers is the steamer *Alonzo*

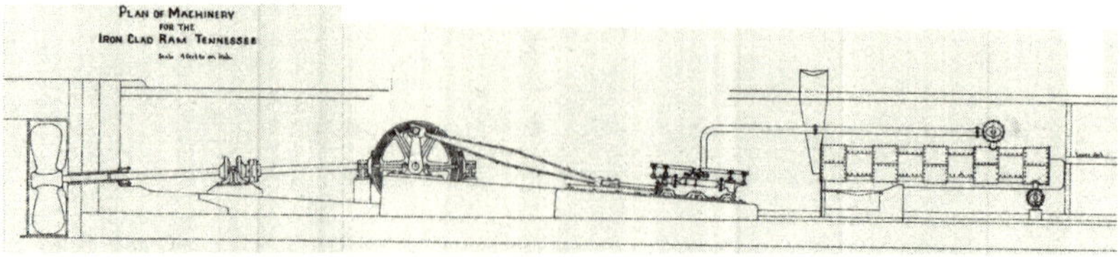

Figure 5.2. Plans of the *Tennessee*'s improvised propulsion system, redrawn by Bob Holcombe in 1982. The original was donated to the Port Columbus Civil War Naval Museum by descendants of Samuel J. Whiteside, a civilian engineer who was involved in the machinery installation. Starting at the right (forward in the ship), the battery of five horizontal high-pressure flue boilers was arranged with the furnace facing aft to consolidate the machinery space into a smaller area. Steam from the steam drum on top of the boiler set is piped into two fore- and aft-mounted inclined paddlewheel engines of the usual riverboat type. Each of these (one on each side of the ship) turns a crank on a large gear wheel through a long connecting rod. The two large gear wheels mesh with and turn two small gear wheels on the centerline (visible to each side of the large wheel) that are connected to the propeller shaft.

Child, which was built in 1857 at Jeffersonville, Indiana, and captured by the Confederates on the Yazoo River.[6] According to a Union survey report made after the *Tennessee*'s capture, the *Alonzo Child*'s engines were transferred to the ironclad.[7] Confusion arises because the dimensions of the *Alonzo Child*'s engines do not match those given for the *Tennessee*, and the time of their removal does not coincide with the installation of the *Tennessee*'s power plant. It appears likely that the engines instead came from the steamer *Vicksburg*. This boat was a 635-ton sidewheeler, 244.5 feet long and 36 feet in beam with a draft of 7.5 feet, built at New Albany, Indiana, in 1857.[8] It was powered by five boilers and two engines, each 24 inches in diameter with a 7-foot stroke. These numbers match those given in the survey report of the *Tennessee*.[9]

The *Vicksburg* was taken into Confederate service as a troop transport but was rammed and sunk by USS *Queen of the West* on February 2, 1863. The Confederates managed to salvage the machinery. The period of the salvage effort appears to fit in the timeline of the *Tennessee*'s construction.[10] It is also stated in a letter from a spy or deserter, a J. J. Brown, to Union Secretary of War Edwin M. Stanton that the sidewheel ironclad CSS *Nashville*, built at the same time as the *Tennessee*, was receiving its machinery from the *Vicksburg*.[11] This is an obvious mistake, because the dimensions of the *Vicksburg*'s engines do not match those of the *Nashville*. In fact, the *Tennessee* and the *Nashville* were confused for each other in several Union reports. In any case, the identity of the *Vicksburg*'s power plant manufacturer is missing from the historical record.

The official Union report on the *Tennessee* described the engines as the typical single-cylinder noncondensing riverboat type with poppet valves, placed fore and aft in the ship (fig. 5.2). In order to make them effectively turn the *Tennessee*'s single 12-feet-in-diameter screw, they were "geared to an idler shaft by spur gearing with wooden teeth, and from the idler shaft to the propeller shaft by bevel cast-iron gear."[12] Overall, it seems to have given remarkably little trouble during the *Tennessee*'s career.

The ship's four boilers were of the horizontal flue type and 24 feet long with a single furnace under all of them.[13] The *Tennessee* was also fairly unusual among Confederate ironclads by being fitted with a mechanical blower (or forced draft fan) to assist in drawing the fires for better steaming. The only others so equipped were CSS *Atlanta* (formerly the English iron hulled blockade-runner *Fingal*), CSS *Huntsville*, CSS *Tuscaloosa*, CSS *Missouri*, and the never completed CSS *Mississippi*.

The *Tennessee*'s propeller, main drive shaft, and probably the custom-made gearing were provided by a source that is well-known, unlike many steamboat machinery manufacturers: the Charlotte Navy Yard. That facility, established in inland North Carolina, was an unlikely source for much of the propulsion machinery produced for the Confederate navy. Founded in 1862 by wartime necessity, the Charlotte Navy Yard ultimately produced the majority of the shafting used by propeller-driven ironclads. It also manufactured ordnance and other naval supplies, having an industrial capacity not easily matched by most other military facilities. This was a result of its close ties to Norfolk before that city's navy yard was recaptured by the Union.[14]

In the spring of 1862 it became apparent that the Confederacy stood a serious chance of losing the Norfolk area and the Gosport Navy Yard. The hotly contested base, where USS *Merrimack* was transformed into CSS *Virginia*, was the largest such facility in the Southern states, and the Union was keen on recapturing it. As the pressure around Norfolk tightened, Captain Sidney Smith Lee, the brother of Robert E. Lee and the yard commandant, was secretly ordered by Secretary Stephen Mallory to begin packing and preparing for transport all the valuable machinery and goods not immediately needed. Beginning in early May 1862, several trainloads of supplies were sent from Norfolk while others were shipped up the James River to Richmond. Captain Richard L. Page, Confederate States Navy, was simultaneously directed by Secretary Mallory to find a safe place in North Carolina for a new facility.[15]

Charlotte was chosen as the ideal location of a new navy yard because it was deep in the interior but still had good railroad connections with the coastal cities. The Charlotte Navy Yard subsequently experienced less disruption by the enemy during the course of the war than any other naval works. The tools and materials shipped to Charlotte from Norfolk were quickly put to use, among them a small steam hammer. A new ordnance works was constructed along with various large structures for the shops, coke ovens, and foundry. The size and output of these facilities increased continually during the war years. By the summer of 1862 a hoisting crane was in operation along with another steam hammer. This new hammer was quite large and was saved from the Union-captured Pensacola Navy Yard in Florida. The tracks of the North Carolina Central Railroad and the South Carolina Railroad passed close by the new Charlotte yard, allowing for simplified loading and unloading of new materials. Overseeing all of this was a force of more than one thousand men led by Captain Page.

Serving under Page, in charge of the large ordnance department, was the able Lieutenant Catesby ap Roger Jones of CSS *Virginia* fame. This man was one of the Confederacy's most skilled naval officers. Jones (whose Welsh ancestry shows in the *ap,* meaning "son of") was another Virginia native, born in Fairfield in April 1821. As a first lieutenant in the Confederate navy, he was the *Virginia*'s executive officer under Franklin Buchanan and became acting captain when Buchanan was wounded. It was Jones who led the *Virginia* into battle against USS *Monitor.* He later commanded the gunboat CSS *Chattahoochee* in Georgia before his brief tenure at Charlotte. Jones finished his wartime career in charge of the Selma Ordnance and Naval Foundry in Alabama, where he had great success in manufacturing Brooke rifles and other weapons.[16]

The Charlotte yard's good fortune in commanders continued after Jones's May 1863 departure; the next commander was Page, who accepted a brigadier's commission in the army.[17] Another skilled veteran of CSS *Virginia* and prominent leader in the construction of the Charlotte Navy Yard, Chief Engineer H. Ashton Ramsay, was next placed in charge. Ramsay was ultimately the commander at Charlotte until the war's end and oversaw the greatest expansion of the navy yard's facilities during that time.[18]

In 1863 the Charlotte yard was further enlarged and improved through the extension and strengthening of its buildings and the addition of new machinery. New pattern shops, a coppersmith shop, a new furnace, and a facility for inspecting projectiles were added at this time. Moreover, the foundry was enlarged to accommodate a new steam hammer built at the yard—the largest in the Confederacy. This piece of equipment was capable of forging the heavy shafting and large anchors required by the ironclads then being built elsewhere. Along with executing all the new developments in manufacturing iron components, the yard continued to turn out gun carriages, blocks, and large amounts of ammunition. Many of these products were destined for service in ironclads as well.[19]

Despite the tremendous capabilities of his establishment, Chief Engineer Ramsay encountered a lack of skilled labor for his workforce. This severely limited production. The great forges and planers of the Charlotte station were too often idle, and many of the heavy forgings intended for new warships took too long to finish. All these symptoms of a small labor pool were encountered by facilities elsewhere in the Confederacy. Attempts by Ramsay to recruit skilled workers from Great Britain failed, and he spent much of 1864 at a minimum level of production.[20]

By the war's end, the Charlotte Navy Yard had nevertheless achieved some notable successes. Thomas Dwyer, one of the workers from Norfolk, invented a machine for turning a perfect sphere, which was particularly applicable to making cannon balls and shells.[21] The machine was first used at Charlotte and was probably the first of its kind anywhere. Large sheds were built for storage of Confederate naval goods, and a variety of large iron shafting for steamers was completed. The ironclads that received their shafting from the Charlotte works were the *Tennessee*, the *Virginia II*, the *Albemarle*, and several unspecified others.[22] Heavy iron components were also provided for several wooden gunboats, and large amounts of ammunition were produced into the war's very last days.

Ultimately, the Charlotte Navy Yard was one of the very last Confederate military establishments to close down in 1865. After the evacuation of Richmond in April, the specie from the Confederate Treasury was sent first to Charlotte for a guard company to take it into the Trans-Mississippi Department. The yard was abandoned and burned not long after, marking the end of Confederate naval administration.[23] Unfortunately, the majority of the yard's records and many of those of the Navy Department itself were burned as well. American history lost a valuable collection of information in that hasty decision.

Back in Alabama, the *Tennessee*'s completion in February 1864 greatly augmented the Confederate naval forces on Mobile Bay. The only ironclad present on that station since the war's early days was the recently condemned *Baltic*. That unwieldy vessel's commander, James D. Johnston, was transferred to become the captain of the *Tennessee*. He served under Admiral Franklin Buchanan, who now placed his flag on board another formidable warship destined to play a prominent role in another famous naval battle.

Johnston was born in Kentucky about 1807, but he spent most of his prewar life in Washington, DC. He served in the US Navy beginning in 1832 but resigned his commission in 1861, initially enlisting in Company B of the Third Alabama Infantry. Johnston next transferred back into naval service and was given the rank of lieutenant, commensurate with his previous experience. As captain of the *Tennessee* he finally had a chance to prove himself as an ironclad captain at the great Battle of Mobile Bay.[24]

Franklin "Old Buck" Buchanan (born September 17, 1800, in Baltimore), the Confederacy's first admiral, remains one of the Civil War's few

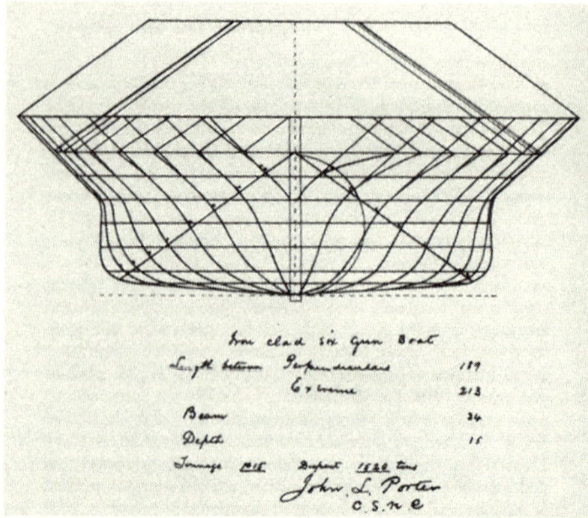

Figure 5.3. Lines of CSS *Columbia*. The writing says, "Iron clad six gun boat, Length between Perpendiculars 189 [feet], Extreme 213, Beam 34, Draft 15, Tonnage 1015, Disp-cmt [Displacement] 1520 tons, John L. Porter, C.S.N.C." Note the improved casemate and knuckle layout superimposed over the old design, as well as the large single propeller in place of the two originally intended. The improved knuckle design increased the ship's sea-keeping abilities because of its above-the-waterline flair, while it also allowed for a wider casemate able to carry more guns on its fore and aft faces.

widely known naval figures. He is best known as the captain (technically flag officer) of CSS *Virginia*, but by the Civil War Buchanan had racked up an impressive record of US naval service since 1815. He was a veteran of the Mexican War and a former commandant of the US Naval Academy. In the Confederate navy Buchanan was initially ranked as a captain but was promoted to admiral after his fierce leadership, gallant conduct, and injury in action on the *Virginia*. Standing exposed on the top deck, he was hit in the leg by sharpshooters on shore. Command accordingly devolved upon Catesby Jones. After his recovery, Buchanan was transferred to Mobile and put in charge of naval defenses. As such he supervised the *Tennessee*'s fitting out and trials. That new and powerful ironclad, a better ship in many ways than the *Virginia*, subsequently became Buchanan's flagship.[25]

Although the *Tennessee* was strongly built and well armored (6 inches of plate on the casemate front, 5 on the sides and aft face), Commander Johnston did not think much of its propulsion. He wrote, "Her speed did not exceed six knots under full steam in slack water. . . . Her engines had been removed from an old Mississippi River steamboat and adapted to a propeller, and its power was totally inadequate to the performance of the work expected of it."[26] Despite this criticism, 6 knots was a respectable speed for any ironclad (even the newest US monitors could manage only about 10 knots), and there was nothing to be done to the machinery installation, anyway, until new engines could be built. Buchanan must have realized as much, because he long shouldered the burden of administering to the *Tennessee*'s construction and fitting out. In the end, Johnston was fortunate that his ship was able to move effectively at all, unlike several other Confederate ironclads.

Buchanan and Johnston did not have long to wait before the *Tennessee* was put to the test. A Union fleet under the command of David Farragut, the conqueror of New Orleans, was gathering outside Mobile Bay throughout the summer of 1864. Farragut was well apprised of the progress on the *Tennessee*'s construction but would not risk an attack without ironclads of his own—namely, monitors. These arrived slowly because of prior commitments, but by August 3, 1864, Farragut's fleet was augmented by the monitors USS *Manhattan*, USS *Chickasaw*, USS *Tecumseh*, and USS *Winnebago*. The federal attempt to take Mobile was set for August 5.[27]

Meanwhile, Buchanan was prepared for the attack but could count on the support of only a few small wooden gunboats. CSS *Baltic* had been stripped of its armor and decommissioned, and the floating batteries *Huntsville* and *Tuscaloosa* lacked sufficient motive power. Therefore the Confederates, with a fairly good ironclad and three armed riverboats, were pitted against fourteen heavy gunboats and sloops of war and four monitors.[28] As the clear morning of August 5, 1864, dawned, the opposing sides began to move.

As the federal fleet steamed into Mobile Bay, its wooden ships formed in a double line ahead with the monitors as a protective screen on their starboard sides, between the fleet and Fort Morgan. At about 6:20 a.m. the battle began with an exchange of shots between the fort and USS *Tecumseh*. Meanwhile, the Confederate ships waited as the Union fleet entered the bay. The *Tennessee* began steering for the lead vessel, the *Tecumseh*, as the fleets came into range of each other. Suddenly a tremendous explosion utterly removed that monitor from the fray: the *Tecumseh* had blundered into the protective Confederate minefield placed at the harbor entrance, sinking within two minutes and taking more than ninety men to the bottom with it. Buchanan next steered his slowly moving *Tennessee* toward the rest of the Union fleet.

Farragut's ships were thrown into confusion by the *Tecumseh*'s sudden destruction. Their order of battle was disrupted, but as each ship came near the *Tennessee* they were still able to outmaneuver the ponderous ram. Nevertheless, Buchanan's flagship poured devastating broadsides into the Union line as it passed. The fighting became general, with the Confederate wooden gunboats engaging their opposite numbers. The *Tennessee* itself continued down the line, receiving several ramming blows from enemy vessels and nearly continuous broadsides. Its iron sides proved their worth: the damage remained minimal. A brief lull then occurred as Buchanan swung around a shoal, which took his ship under the protective guns of Fort Morgan.[29] Despite his mighty efforts, the enemy fleet was now completely inside Mobile Bay.[30]

Instead of waiting to resume the fight under cover of darkness, Buchanan again took the *Tennessee* out after a brief respite. This time he steamed directly into the enemy's midst and engaged the entire fleet singlehandedly. The Confederate admiral's likely purpose in this seemingly suicidal attempt was to strike a decisive blow before Farragut had time to recover and consolidate his forces. The *Tennessee* was soon completely surrounded by the Union ships. Continuous broadsides and more ramming attempts still could not disable the Confederate flagship, which continued inflicting tremendous damage to its wooden adversaries. It is remarkable that the *Tennessee*'s improvised power plant apparently did not malfunction under such a tremendous beating, but the ship's smokestack was completely riddled and soon collapsed, greatly lessening the draft to the boiler fires. As choking fumes permeated the already smoke-filled casemate, the *Tennessee* began to lose steerageway. It still appeared as if Buchanan could hold out until the federal monitors at last entered the melee. Almost immediately a solid 15-inch shot from USS *Manhattan* partly penetrated the *Tennessee*'s side. The other monitors began hammering away from every angle. Admiral Buchanan himself was wounded, and soon the steering chains were shot away, sending the flagship drifting out of control. It became apparent that the only choices left were to go to the bottom or strike the colors (i.e., lower the flag to surrender). With only half his guns still operational, Buchanan gave Commander Johnston permission to surrender.[31]

Thus the Union navy captured Mobile Bay and one of the Confederacy's most powerful ironclads. The federal forces still had much work to do to gain the city of Mobile (where three more Confederate ironclads—the *Tuscaloosa*, the *Huntsville*, and the incomplete *Nashville*—remained) but Farragut had scored another dramatic victory. The Confederate naval force that had stood against him was utterly defeated. Two of the three wooden gunboats were destroyed, and Buchanan, Johnston, and the remaining crew of the *Tennessee* were captured. The *Tennessee* itself was surveyed, extensively repaired, and placed in Union service. It was sent to New Orleans and spent the rest of its career as part of the US Mississippi Squadron.

After its wartime service the *Tennessee* was laid up and sold for scrap on November 27, 1867.[32] Because of this (unlike most Confederate ironclads), there are no material remains for determining why its makeshift propulsion system worked so well. It certainly represented a very rare, moderately successful case of powering an ironclad. The old steamboat *Vicksburg*'s paddlewheel engines ultimately saw the *Tennessee* through one of the Civil War's greatest naval engagements.

CSS *Columbia*

CSS *Columbia* was the much-modified sister ship of the *Tennessee* and one of the most powerful vessels commissioned by the Confederates, with an armament of four pivot and two broadside guns protected by 6 inches of armor.[33] This vessel was the last ironclad completed at Charleston, but it never had a chance to be proven. The *Columbia*'s career was over almost as soon as it began, cut disastrously short because of human error.

This ironclad was built by Francis M. Jones at Charleston—laid down in December 1862 and launched on March 10, 1864.[34] The *Columbia* was modified in the course of construction to a refined knuckle shape: instead of sloping back from the waterline as on other ironclads, the new design continued to flare up to main deck level (fig. 5.3).[35] This improved layout increased seaworthiness and gun deck room, allowing more pivot guns to be carried.[36] The *Columbia* was also originally designed to be driven by two propellers but was finished with a single three-bladed screw 10 feet 8 inches in diameter turning on an 8-inch-diameter shaft.[37] All these design changes severely delayed construction (along with the omnipresent material and labor shortages encountered on virtually all other late-war ironclad construction projects), but the end product was an impressive warship. When completed, the *Columbia* measured 215.78 feet long and 48.5 feet in beam with a draft of 13.5 feet.[38]

The *Columbia*'s power plant was brand-new and manufactured specifically for the ship by the Columbus Naval Iron Works.[39] Five horizontal five-flued boilers—each 22 feet long and 4 feet wide, for a total grate area of 129 square feet, with a lower flue diameter of 12 inches and an upper one of 10.5 inches—powered two noncondensing horizontal single-cylinder engines, each 36 inches in diameter and with a 24-inch stroke.[40] The propeller and shafting were also probably forged at Columbus. This system promised good performance and actually represented one of the Confederacy's finest engineering accomplishments.[41]

The *Columbia* was never effectively tested in service, for just days after its commissioning in January 1865 the ship ran hard aground on a sunken wreck in Charleston Harbor.[42] A firsthand account of the disaster is provided by the ubiquitous Eugenius A. Jack, who was transferred to Charleston after his service at Wilmington. Jack was also promoted to first assistant engineer by this time, and he reported on board the still not-quite-finished *Columbia* as an aide to the Charleston fleet engineer, Virginius Newton Freeman.[43]

Freeman was a native Virginian, a former resident of Norfolk like so many other Confederate naval engineers. He was born about 1834, but despite his relatively young age he was ranked as a first assistant engineer in the US Navy when he resigned in 1861. Freeman was commissioned into the Confederate navy in July 1861 with the rank of chief engineer, and he served at New Orleans until that city fell in April 1862. He was then at Jackson, Mississippi, before being transferred to Charleston as the senior engineer of the squadron there. Freeman's prewar experience must have been considerable, and his wartime appointments quite successful, for him to rise so quickly in the steam engineering hierarchy.[44]

First Assistant Jack, for one, did not think too highly of his senior engineer. Upon arriving in Charleston he stated the following:

> This vessel [*Columbia*] was not yet compleated [*sic*]. . . . So with the other officers ordered to her I was assigned quarters on the Receiving Ship at that station. I cannot recall her name—it was with no little pleasure that I left the crowded quarters there for the pleasant cleanly ones of the new ship a short time before she was ready for the trial trip. But although living aboard the ship, I was in no way responsible for the machinery as yet, for it was still in the hands of the fleet-Engineer Mr. Virginius Freeman, and would not be placed in my charge until after the trial trip. Bear this in mind, for I am about to tell you of a shabby trick that officer attempted.[45]

Jack was apparently slated to become the *Columbia*'s acting chief engineer after the new ironclad's trials. He had certainly earned it. He was impatient for the successful completion of the trials, but shortly everything began to go wrong.

During the trial the *Columbia* steamed out into Charleston Harbor at high tide. The course took the ship over some sunken obstructions navigable only at high water. Jack recounted the trip and what happened next:

> At last the engines and boilers were ready for the trial, and there was a gathering of officers from other vessels of the fleet. Steam was gotten up by the force sent from Columbus, Ga. by Chief Engr. Warner, who had constructed the machinery at the Govt. works at that place, and his men were at the engines. Mr. Freeman, as Fleet Engineer, was the supervising officer, and I a looker on and, to some extent, an assistant to him also. There was [*sic*] some drinkables and eatables on the spread in the wardroom, too much of the drinkables I fear. Not long after we got underway, a valve was opened by one of the shop hands, which came near to causing the loss if not the explosion of one set of the boilers. The fires under these had to be hauled [put out], because they were nearly empty of water [presumably drained

> by the opened valve]. After the fires were hauled and the boilers were cooled they were refilled and fire started again. When this bad piece of engineering occurred, the ship was steaming to reach a shoal near the mouth of the stream while there was enough water to pass over. The loss of steam and slow consequent speed of the engines caused her to miss the tide, and when on the way back to her moorings, she was run upon a sunken vessel, and hanging there, was left by the tide, and leaking in her hull, became a wreck.[46]

The damage must have appeared catastrophic, for the Confederates abandoned the ship, leaving it where it was and stripping off the most valuable materials over the next few days.

Shortly afterward, Chief Engineer Freeman turned on First Assistant Jack. According to the latter, "Now when the board of inquiry [on *Columbia*'s loss] met, Mr. Freeman, instead of stating this serious delay that was caused by arrant carelessness, mentioned a slight delay that was caused by a heated journal [bearing] which I had set down upon at his discretion and under his supervision. At most, the delay was not three minutes from this cause, and it was as much his fault as mine, and again a thing that might occur with the most experienced engineer, with new machinery. His intention was to make me the Scape Goat [*sic*]."[47]

Jack was not called upon to testify, but he learned from his friends what was said at the inquiry. Despite Freeman's attempt to place the blame on the junior engineer, neither of them seems to have been punished—at least Jack makes no mention of it. He was detached to the *Columbia*'s squadron mate, the *Palmetto State*, after a few days, where he remained until Charleston was evacuated and the ironclads destroyed.[48]

It can be readily seen that even the most experienced steam engineers could make catastrophic mistakes, especially when they were supervising machinery they were not familiar with. In this case it cost the Confederacy one of its best-built and most powerful ironclads. The damage was apparently not as severe as originally thought, however. After the fall of Charleston in February 1865 the US Navy made a survey of the wrecked *Columbia* and ultimately succeeded in refloating it. The valuable hull was actually moved out of Charleston Harbor under its own power and taken to Norfolk.[49] The *Columbia* was not returned to service and was eventually sold for scrap. Thus neither side benefited in any appreciable way from the service of one of the Confederate naval industry's finest products.

Conclusion

The *Tennessee* and the *Columbia* ultimately represented the rapid pace of improvement in both Confederate ironclad design and steam power.

Though technically sister ships with the same standard hull design, the end products were very different. Each was successful, although the *Tennessee* had serious weaknesses and the *Columbia* was never truly put to the test.

The *Tennessee* still suffered from secondhand machinery (in an especially unusual arrangement) but proved a powerful warship and was vastly improved over earlier designs, even Admiral Buchanan's former command CSS *Virginia*. Its machinery worked surprisingly well, validating in many ways the use of high-pressure systems removed from riverboats. An increasing number of late-war ironclads used these systems simply because of their ready availability and relatively high level of power generation for their size. The *Tennessee* set the precedent for this in many respects, although it remained the only Confederate ironclad that successfully utilized riverboat machinery to turn a propeller. Therefore, the *Tennessee*'s unorthodox installation can be considered somewhat of a success, in both its Confederate and its federal service. The Confederates lost the ship primarily because of the overwhelming number of opponents in Mobile Bay, not through any critical mechanical breakdown.

The *Columbia* was a far different story. This ironclad featured the latest design in improved seakeeping, gun deck layout, and reduced draft coupled with brand-new machinery from Columbus. Because of these fine qualities it had more in common with the last ironclads built during the Civil War than with the *Tennessee* or even most of the other standard hull vessels. The Confederates suffered a great loss when the *Columbia* was wrecked so soon after completion, but it is difficult to say what (if any) effect the ship's presence might have had on the final defense of Charleston.

The engineers who installed or built both vessels' power plants continued to represent the rapid maturation of the Confederate wartime industry. The *Tennessee*'s improvised setup worked, and the *Columbia*'s new one was a fine mechanical achievement. The engineers on each ship in service were also for the most part highly competent, but the *Columbia*'s loss provided a rare and dramatic example of how a small slipup (or slight state of intoxication) could be disastrous. Even a brand-new steam warship of improved design could not function without an engineer's constant vigilance and familiarity.

The improvements embodied by the *Columbia* were later applied to other Confederate ironclads designed by Constructor Porter (in most cases the changes were made during construction), but in the meantime some other interesting types of standard hull ironclads were built. Among them were two excellent vessels not designed or built by Porter (unusual in itself) and one poor one of questionable design and construction.

SIX

OTHER STANDARD HULLS

Mr. Graves vessel, to be iron clad . . . will be . . . strong and fine.
—Lieutenant John M. Brooke, Confederate States Navy

The last group of standard hull ironclads built by the Confederacy presented even more design differences and unusual features than the *Tennessee* class ships. All were large like their predecessors, but there most similarities ended. The *Charleston*, the *Virginia II*, and the *Nashville* differed substantially in ways other than the hull design. These ironclads (possibly excepting the *Virginia II*) are relatively little documented. Many details of their steam machinery remain an enigma.

The *Charleston* and the *Virginia II* can be considered sisters or near sisters, since both were designed and built by a naval constructor other than the prolific John L. Porter, who conceived most of the Confederacy's ironclad designs. These vessels accordingly had some important differences that reflected both wartime experience and the ideas of their builder, Acting Naval Constructor William Graves.

William Armistead Graves (born September 4, 1821) was, like Porter, a native of Norfolk. He was a shipwright before the war but enlisted in Company C, Ninth Virginia Infantry, in September 1861. Graves's valuable skills were soon recognized, however, and he was appointed acting naval constructor on March 22, 1862. He spent the war at Richmond and even had his own shipyard there (known only as "Graves's Yard"), where the *Virginia II* was built. Graves subsequently became the only other major contributor to Confederate ironclad design besides Naval Constructor Porter, and his two completed vessels saw some success. After the war Graves returned to Portsmouth and resumed his career as a shipbuilder.[1]

The *Charleston* and the *Virginia II* reflected continuing changes in Confederate ironclad design over the period of their construction. As with the *Tennessee* and the *Columbia*, the finished products were quite different from each other because of where they were built, the availability of local materials, lessons learned from other ironclads in combat, and the different times at which construction began on the two vessels. The *Virginia II* was finished last and incorporated significant design changes, making it probably the strongest ironclad completed by the Confederacy.

The *Nashville*, which was not designed by Graves, is included in this group because so little is known about its hull configuration. It and its uncompleted sisters are generally assumed to be standard hulls. The *Nashville* was a rare case in which the ship's construction itself, not the machinery, was the primary cause of weakness. In fact, the machinery sets on the *Nashville*, the *Virginia II*, and the *Charleston* seem to have performed fairly well overall.

CSS *Charleston*

The third ironclad built at Charleston, South Carolina, was named in honor of that city. CSS *Charleston* was built by the firm of J. M. Eason & Brother, which began construction in December 1862.[2] This ironclad was the first one built to Graves's design and is notable chiefly for its flat floors and soft bilge turns, compared to Porter's vessels, as well as an unusually long protruding ram. The latter feature was the Graves ironclads' most distinguishing feature, but its design remains mysterious. Possibly it was intended for use against monitors, whose armored rafts protected their hulls much as the knuckles of Confederate ironclads protected theirs. The long extension and curved upper surface of the Graves ram may have been intended to slip under the raft and strike a monitor's unprotected hull.[3]

Previous wartime experiences of ramming may also have been kept in mind. The Graves ram's somewhat fragile shape, with a thickness of only one foot, could have been designed to break off after insertion. The *Virginia*'s experience after ramming USS *Cumberland*, when only the breaking of the ram prevented the *Virginia* from being dragged to the bottom, clearly demonstrated the danger if the weapon was not quickly withdrawn.[4]

Although both hypotheses are plausible, neither the *Charleston* nor the *Virginia II* ever had the chance to ram an enemy vessel. Their design (at least as Graves originally conceived it) was otherwise fairly standard. This is known through the preservation of two half-hull models in the US Naval Academy Collection at the Smithsonian Institution.[5]

As with most Confederate ironclads, changes were made to the *Charleston* during construction. The ship was modified to carry two pilothouses (one forward and one aft), and the casemate was changed to a simpler four-sided layout instead of the original octagonal. In addition, the upper bow was modified for carrying a spar torpedo. The end product was a large ironclad measuring 180 feet long between perpendiculars, 201 feet long overall, 44 feet in extreme beam, 14 feet in depth of hold, and a little more than 12 feet in draft.[6]

The *Charleston* was finished in early 1863 and commissioned on August 19 of that year. A single 9-feet-in-diameter screw powered by one or two direct-acting low-pressure engines pushed the *Charleston* along at approximately 6 miles per hour. Nothing is known about the boilers. All the information presented here comes from an August 1863 entry in the John Horry Dent Papers, a restricted collection kept by the University of Alabama. No other information on the *Charleston*'s machinery or its source has been found. It is possible that the shipbuilder, the Eason foundry, manufactured it, since that establishment certainly had the experience and the capability.

Isaac Newton Brown of CSS *Arkansas* fame became the *Charleston*'s commander, but by the time the ship was commissioned the major local naval events in which ironclads participated were over. The *Chicora* and the *Palmetto State*'s dramatic but ineffective raid had occurred almost half a year earlier, and the one major federal assault on Charleston Harbor by water ended in repulse by the Confederate forts. Brown accordingly had little to do with his powerful new command during its career. The *Charleston* became the flagship of the squadron, but as with the *Chicora* and the *Palmetto State*, its primary duties consisted of steaming down to Fort Sumter on alternate nights to act as a guard ship. The flagship spent its entire career in such relative inaction.[7]

In the end, the *Charleston* was blown up along with the other ironclads to prevent capture, when its namesake city fell in February 1865. Its pieces lie in dark, polluted, and heavily traveled waters. A dedicated investigation of any remaining wreckage may uncover more details on its mysterious machinery installation.

CSS *Virginia II*

The second Confederate ironclad named the *Virginia* was quite similar to the *Charleston* as originally designed by Graves. The final product took longer to complete but, as modified, represented probably the strongest ironclad commissioned by the Confederacy. CSS *Virginia II* featured a shorter, stronger octagonal casemate with four guns and additional armor: 6 inches in thickness on the sides and 8 on the ends.[8] In addition, cambered and sloped flush decks just forward and aft of the casemate were installed. These provided more ablative and angled armored surfaces for projectiles to glance off.

The *Virginia II* was built at Graves's Yard directly across the James River from Rocketts, east of downtown Richmond. Constructor Graves therefore had direct control over construction, from keel laying to launch on June 29, 1863.[9] The finished ironclad differed slightly from its predecessor the *Charleston* in basic measurements: 203 feet long overall, 48.5

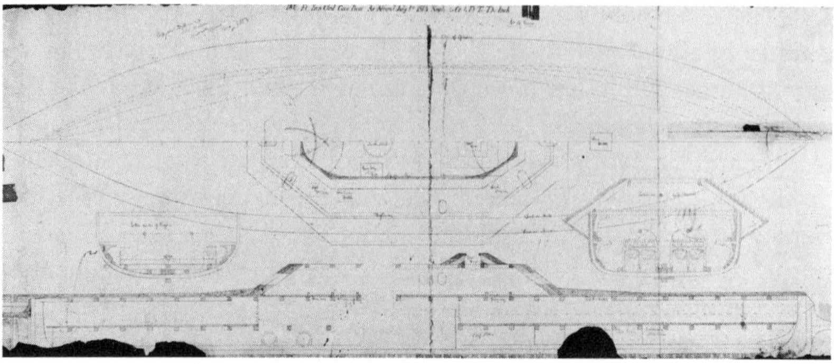

Figure 6.1. 1863 plans of the *Virginia II*, showing the machinery layout in addition to the hull construction. It is not known if this view represents the final configuration of the engines. They are represented in a horizontal layout, as seen in the engine room cross-section at the left, between the profile and plan views. Overhead valve chests like those on CSS *Raleigh*'s engines can be seen. The propeller shaft is shown at the center, and the aft side of the boilers (with main steam lines) is represented by dash lines. No linkages between the engine cylinders and propeller shaft are shown. In the midships cross-section at the right, the forward ends of the boilers are shown and are revealed to be of the standard horizontal firetube type. They were placed backward in the hull to consolidate the machinery spaces, so their furnaces were facing the engines; this is revealed by the location of the ship's smokestack. Note the heavy construction of the short casemate and the angled fore and aft flush decks.

feet in beam, and a 14-foot draft.[10] Its hull form remained the same, but there was an apparent machinery improvement: the *Virginia II* was supposedly capable of up to 10 knots with its possibly English-built engines.[11]

As discussed in an earlier chapter, there is some uncertainty about the provenance and final layout of the *Virginia II*'s power plant. Several records suggest that it was built new either at Tredegar Iron Works or the Shockoe Foundry (Confederate Naval Works) just a few miles away from Graves's Yard.[12] Tredegar was at least in charge of fitting out the ironclads in Richmond. Confusion arises because at least one account suggests that the *Virginia II* may have had British machinery of unknown origin.[13] In an April 25, 1865, survey of captured steam machinery, Commander William Radford, US Navy, listed "one pair of double engines, 30-inch cylinder and 28-inch stroke, high pressure and similar to those used on Ram Virginia [II]."[14] *Double engine,* in nineteenth-century parlance, generally meant a two-cylinder engine; Great Britain was the primary manufacturer of twin-cylinder engines by the 1860s. Therefore it appears likely that if the *Virginia II* did have the type described by Radford, they were made by a British or Scottish manufacturer. Scottish industry particularly was the world leader in iron construction for both hulls and engines. Twin-cylinder engines, both simple and compound, were successfully employed by them long before the type was common in the United States.

Fortunately for the modern researcher, some information can be gleaned from the surviving plans of the *Virginia II* (fig. 6.1).[15] The plans seem to have been made after major modifications to the ship were completed or at least approved. They show all the *Virginia II*'s notable features but most important the machinery, in some detail. This is relatively rare for extant plans of Confederate ironclads.

Although it remains unknown whether the plans represent the final machinery layout, they show two single-cylinder horizontal direct-acting engines approximately 40 inches in diameter and 38 inches in stroke that were of the typical design manufactured for several ironclads. If we assume that the plans can be trusted in regard to boiler if not engine layout, the *Virginia II* was powered by two horizontal boilers with two furnaces each. The boilers were 22 feet long, 8 feet in diameter, and 9 feet high. Propulsion was provided by a single 10-feet-in-diameter screw turning at the end of a propeller shaft 54 feet long and 11 inches in diameter.

A few bits of firsthand engineering testimony also exist. Three very brief but informative statements concerning the *Virginia II*'s steam machinery in service are located in the US Navy's *Official Records of the Union and Confederate Navies in the War of the Rebellion*. They range over the course of a year, between February 1864 and February 1865. Evidence of the hasty completion of the ship can be seen, but there are no significant details on the layout of the engines themselves. The only hint in this direction is in an informal letter written on February 26, 1864, before the ship's completion: "She has an engine of nearly 500 horses."[16]

The *Virginia II* became the James River Squadron flagship when it joined the fleet in May 1864. The new ironclad's captain, transferred from CSS *Richmond*, was the veteran Robert Pegram.[17] He was joined by Flag Officer John Mitchell, and the *Virginia II* was soon undergoing trials. All was routine for about a month. Then on June 21 Mitchell telegraphed Secretary Mallory about a problem eerily similar to that encountered by CSS *Savannah* nearly a year earlier: "I was about proceeding down the river, near Howlett's [a Confederate land battery], when the piston was discovered to be out of order. The cylinder was removed and a chisel found in the cylinder. The engine is now in working order again."[18] The damage must have been minimal for the repairs to have been made so quickly. The incident again showed the problems inherent in hasty construction and the general lack of experience many workers had with steam machinery.

Mitchell was better off than many commanders, however. Under the James River Squadron's capable chief engineer, Henry X. Wright, weekly machinery inspections were instituted. On July 7, 1864, Wright finished looking over the new flagship. He wrote to Flag Officer Mitchell, "Sir: I have the honor to report that at general inspection held on board the C.S.S. Virginia [II] this day, that after a very careful and thorough examination of the engines, boilers, steam pumps, and the appendages, under steam, I found everything in good order and condition and in a very high state of efficiency."[19] With this blessing the *Virginia II* presumably ended its trials and became a fully operational unit. Unfortunately there was little to do. The flagship took part in the James River Squadron's one major engagement, the Battle of Trent's Reach in January 1865, but achieved little. The *Virginia II* grounded several times because of its draft and became the Union gunners' largest and easiest target. Mitchell was able to get the ship refloated but not before several heavy hits were scored, particularly from the large monitor USS *Onondaga*.[20] Fortunately for the Confederates, the *Virginia II* was so heavily armored that the *Onondaga*'s 15-inch shots could not quite penetrate.

Back at its usual mooring area upstream, the squadron underwent repairs. Each ship was brought back up to peak condition (at least as far as the available materials allowed). During this time a major safety problem was found with the *Virginia II*'s auxiliaries. The temporary commander of the ship in February 1865, Lieutenant John Dunnington, wrote on January 5, "On berth deck the force pumps, intended to use in case of fire or when the crew could not go outside of shield, have never been completed."[21] It is unknown whether the pumps were ever completed, but they were never put to the test. The *Virginia II* and its squadron mates had only another month of service. During that time they did not venture out again.

Near the war's end a good account of the interior layout of a Confederate ironclad was written by a visiting Irishman, Thomas Conolly, who toured the *Virginia II* in late March 1865. This was just a week before the evacuation of Richmond on April 2–3.[22] Conolly's account is revealing because it makes clear just how modern and well maintained many Civil War ironclads were compared to other vessels, despite their engineering and climate control problems (unfortunately, no detail is given on the *Virginia II*'s machinery spaces):

> [We went] thro a small iron port hole where we saw the thickness of her iron-sides lined with oak about 4 feet [thick] when inside the great guns one astern of gigantic proportion d[itt]o for'ad & 3 [gunports] at either side light & air coming thro thick iron bars at top sloping inwards garnished with all the gun requisites tower tier filled with long steel soled shot about 12 inches each. Middle supports covered with racks of six & ten shotting rifles in fine order men & officers sauntering about quietly—Thence to mens quarters down a hatchway for'ad all in order with a cooking stove full of excellent dinner surrounded by mess boxes wh[ich] when opened had their tin plates,

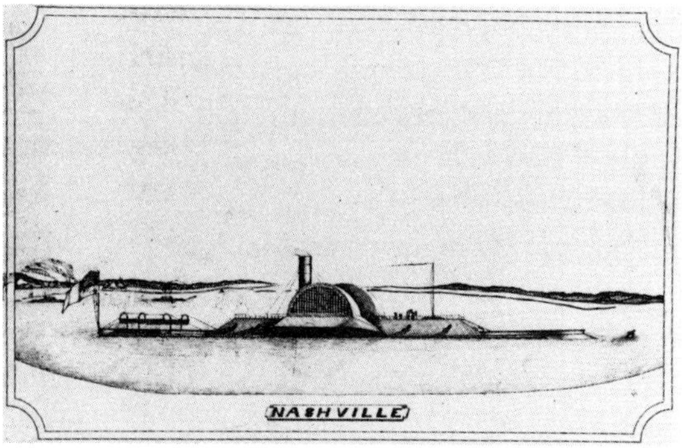

Figure 6.2. Contemporary drawing, circa 1865, of the large sidewheel ironclad *Nashville*. The massive paddlewheel boxes dominate the ship's profile and indeed appear a select target for naval gunners. Despite its cumbersome appearance, the *Nashville* apparently handled quite well.

> cups knives salt pepper &c in good array with the n[umber] of rations for each mess in middle! All as nice as any man of war hammocks all stowed along inside making a pleasant couch to lean against— Men all standing round at attention & all neatly clad in confed: grey shirts. Then to engine room at other end—Thro wardroom & quarters Ad(mirals) sleeping room all dark but bright & clean when lighted up.[23]

Conolly was one of the last civilians to see the Confederacy's only remaining naval squadron intact. Mere days after his visit, the city of Richmond was evacuated by the Confederate government. The city went up in flames as government workers, the military, and civilians alike abandoned the capital. There was now no purpose for the James River Squadron. Union forces were ascendant below its position and soon to be above it. The squadron, now under the command of Raphael Semmes of CSS *Alabama* fame, was slated for self destruction. The *Virginia II* therefore suffered the same fate as most other Confederate ironclads. The strongly built and powerful ship was blown up (along with the remaining vessels of the James River Squadron) in the early hours of April 3, 1865, to prevent capture. It is uncertain what, if anything, remains of the wreck today.

CSS *Nashville*

CSS *Nashville* was the only Confederate ironclad completed from the keel up to take advantage of the availability of sidewheel riverboat machinery. (In contrast to the *Tennessee's engines*, the *Nashville*'s engines served their original purpose in turning wheels.) The *Nashville* remained a unique vessel, although contracts for four similar ironclads were originally drawn up.[24] Contemporary sketches reveal the *Nashville*'s appearance (fig. 6.2), but no builders' drafts are known to exist, and the lower hull form is uncertain. The overall characteristics and measurements seem to point to a standard hull configuration. Fortunately, a Union survey of the captured *Nashville* provides significant details on its construction and machinery layout.

The ironclad was constructed at Montgomery, Alabama, by J. E. Montgomery and A. Anderson and was one of the largest in the Confederacy. The *Nashville* measured 271 feet long overall with a hull beam of 62 feet, a beam over the paddleboxes of 95.5 feet, a 13-foot depth of hold, and a draft of 10 feet 9 inches.[25] It was taken downriver after its launch in May or June 1863 for completion at Mobile, but the usual lack of materials and labor severely delayed progress.[26] Not least among these problems was an almost total lack of iron plating for armor and the new ship's increasingly apparent structural weakness as a result of sheer size. Enough armor to cover the fore and aft casemate ends and the pilothouse was obtained from the rotten *Baltic*, but even with this light covering, the extensive use of hogging trusses and chains, and a reduced armament of three guns, the *Nashville* continued to hog.[27]

Despite the weight issues, it seems that the vessel's engine performance was better than that of many of the screw-propelled ironclads. The giant paddlewheels were greatly exposed to combat damage, but the Navy Department seems to have believed that the ease with which high-pressure, long-stroke engines could be installed, with few if any modifications, outweighed any disadvantages.[28] As previously discussed, machinery from captured, purchased, or wrecked riverboats was the Confederacy's one nearly universal source of ready-made engines and boilers. Unlike many other ironclads utilizing these systems, the *Nashville* was fully intended to exploit steamboat machinery with a minimum of adaptation. The Confederates might have been better off employing such a design philosophy (for river ironclads, at least) from the beginning.

There is some uncertainty about the source of the *Nashville*'s machinery, largely because of the confusion in Union reports between this ironclad and the propeller-driven *Tennessee*. At least two reports state that the river steamer *Vicksburg*'s engines were used for the *Nashville*.[29] It has already been seen that the given dimensions of the *Vicksburg*'s engines and the time of their removal fit with the available data for the *Tennessee*. Fortunately, evidence also exists for the source of the *Nashville*'s engines in yet another Union statement that confuses the *Nashville* with the *Tennessee*. That report suggests they came from a wrecked Yazoo River steamboat.[30] The only vessel in that river with machinery salvaged by the Confederates and with matching engine dimensions was the large sidewheeler *Magenta*, taken up the Yazoo and burned on July 14, 1863.[31]

The boat was a wooden packet of 782 tons built at New Albany, Indiana, in 1861. Its hull was constructed by Dowerman & Humphreys, and the machinery by Lent, South & Shipman. The *Magenta* was 285 feet long, 40 feet in beam, and 8.5 feet in depth of hold, while its sidewheels, which propelled the boat, were 37 feet in diameter and 16 feet wide.[32] The dimensions of the vessel's eight boilers fit with those on the *Nashville*.

The only other likely candidate for the source of the *Nashville*'s machinery is the steamer *Capitol*. Its stated machinery dimensions also match the ironclad's. The *Capitol* was a fast sidewheel packet built in 1854 at Jeffersonville, Indiana. The vessel measured 224 feet long, 32 feet wide, and 6 feet in draft. The *Capitol* gained fame for its speed and for aiding in the completion of CSS *Arkansas* at Yazoo City, Mississippi. The boat's machinery was used in completing the ram, and the *Capitol*'s hoisting engines were made over into drilling machines and other shipworking tools.[33] The *Capitol* was later burned to prevent capture at Liverpool, Mississippi, on June 28, 1862. All the ironwork and machinery was successfully removed by the Confederates.[34] The provenance of the *Capitol*'s engines and boilers is currently unknown.

The *Nashville*'s machinery dimensions (if not the source) are so well-known because the ship became one of the few Confederate ironclads to fall into enemy hands. Two US Navy reports, one by Rear Admiral Henry K. Thatcher and another by Acting Volunteer Lieutenant George P. Lloyd, described the *Nashville*'s power plant. There are some minor discrepancies: Thatcher listed the engines (one per wheel) as 30 inches in diameter with a 9-foot stroke, whereas Lloyd listed the cylinder diameter as 32 inches.[35] These numbers may represent the bore diameter versus the outside diameter of the engine cylinders. The *Nashville*'s seven boilers were reported as the double-flue type, 30 feet long and either 40 inches (as reported by Thatcher) or 42 inches in diameter (as reported by Lloyd). Lloyd's report also listed the paddlewheels as 28 feet in diameter and 12 feet wide.[36]

Whatever its source, the *Nashville*'s powerful machinery was successfully installed at Mobile. When Lieutenant Charles Simms was appointed the new ironclad's captain on July 21, 1864, it was functional but still incomplete. Armor was so difficult to obtain that it was decided to remove all plate from Simms's previous ship, CSS *Baltic*. Ultimately there was only enough plate to cover the forward end of the casemate and the pilothouse with three layers of 2-inch plate and to put one layer on the aft end. At some point before its capture, the *Nashville* was also fitted with hogging braces and chains in an attempt to counteract the apparent weakness of the giant hull. Worry over structural strength combined with a lack of available ordnance also meant that a reduced armament of only three main guns (all 7-inch Brooke rifles) was carried. This was the ironclad that Simms moved to, although it was still better than the *Baltic*. The *Nashville* was never really finished, however.[37]

The *Nashville* was still working up when Franklin Buchanan took CSS *Tennessee* and the gunboats of the Mobile Squadron out to fight David Farragut at the Battle of Mobile Bay. Although the federal fleet successfully beat the Confederates and entered the bay, the city of Mobile remained untaken. The Union drive from both land and sea to capture the city lasted the rest of the Civil War. The main line of defense was Spanish Fort on the Blakely River, where the *Nashville* participated in one of the Confederacy's last stands.[38]

After some months of idleness, while the Union forces consolidated their positions on the lower bay, Lieutenant Simms was detached from the *Nashville* and transferred to Selma Ordnance and Naval Foundry. Once again he missed a chance to lead an ironclad into battle. By this time it was late November 1864. The *Nashville*'s new captain was Lieutenant John W. Bennett (born in Maryland in 1822), a naval veteran of the Mexican War. He was also an 1846 graduate of the US Naval Academy but resigned from the US Navy in April 1861. Bennett spent most of his Confederate service on the Mobile station before taking over command of the *Nashville*. In March 1865 he had the chance to test the ship in combat.[39]

By March 27 the Union was severely punishing the Confederates at Spanish Fort in repeated attacks. The *Nashville* was ordered to support the fort by taking up a position where it could shell the Union land forces. Bennett took his ironclad downriver; although the huge ship steamed well, some trouble with deep draft was again encountered when the *Nashville* grounded for two days. Finally the Confederates were able to refloat it with minimal damage, and the *Nashville* took up its appointed position shelling Union troops and using its boats to remove Confederate wounded. A brief duel with a shore battery resulted in some damage to the aft casemate and rear pivot gun, but the ironclad remained in its supporting role until April 1, when Bennett steamed his command back to Mobile.[40]

The *Nashville* proved once again the effectiveness of naval support for ground operations, but shortly after, the Confederates were forced to give up the untenable defense of Spanish Fort. Union pressure on the area grew daily. The *Nashville* continued to steam back and forth during the early days of April 1865, harassing and shelling the federal forces, usually to cover Confederate retreats. Ammunition supplies became very low. Everywhere the Confederacy was on its last legs. Finally, on April 12 (three days after Appomattox), Mobile surrendered. The *Nashville* and the remaining Confederate naval units were now without a port. Under the overall command of Captain Ebenezer Farrand, the little force steamed up the Tombigbee River to make a last stand. It never happened. Probably realizing the futility of such an effort, Farrand opened negotiations with Union Rear Admiral Thatcher; they agreed to May 10, 1865, as the surrender date. On that day at Nanna Hubba Bluff, 40 miles from Mobile, the last Confederate fleet surrendered. The US Navy obtained the *Nashville* and its consorts: the gunboat *Morgan*, the ex-ironclad *Baltic*, and the steamboats *Black Diamond* and *Southern Republic*, as well as 112 officers, 285 men, and 24 marines.[41]

Lieutenant Bennett's *Nashville* was the greatest captured prize at Nanna Hubba Bluff. Union officers surveyed the ironclad, and it was taken to New Orleans. By that time, however, no more warships were needed. The *Nashville* remained incomplete (its valuable armor was stripped off for other uses) and was laid up. The giant vessel was sold in November 1867.[42]

Conclusion

The last group of standard hull ironclads reflected significant design changes from the earlier *Richmond* and *Tennessee* class vessels while keeping the same basic hull form. The *Charleston*, the *Virginia II*, and the *Nashville* were also quite different from one another when completed. They served in widely scattered areas of operation, and the *Charleston* and *Virginia II*'s builder, Constructor William Graves, conceived some unique features (such as the unusual ram) that distinguished his vessels. Graves ultimately became the only other major contributor to Confederate ironclad design besides Constructor John Porter. The *Charleston* and the *Virginia II* were strong vessels, and their machinery installations seem to have been acceptable. It is unfortunate that there are so few firsthand engineering details for these ships.

Although it is still difficult to say for certain that the *Nashville* was a standard hull ironclad, its dimensions as given in the Union survey and its known above-water configuration suggest the standard hull form. The *Nashville* was unique in using sidewheel riverboat machinery (the only Confederate ironclad to do so besides the converted *Baltic*). This system, though exposed to combat damage, was effective because such power plants were readily available and powerful for their size. The *Nashville* had many problems with weight and a lack of construction materials, but its engines seem to have functioned well despite the size of the ironclad.

Very little primary source material has been found concerning the steam machinery of these three ironclads. This is particularly unfortunate, because in many ways the *Charleston*, the *Virginia II*, and the *Nashville* represented the penultimate stage of Confederate ironclad design.

All saw service, and the latter two some amount of combat, unlike the much-improved late-war designs (all unfinished; see chapter 9). The *Virginia II* especially marked a peak in resistive strength, with multiple angled surfaces and 8-inch thicknesses of armor. It is hoped that more information, disclosing at least the manufacturers of these ironclads' power plants, will be discovered.

Despite the strengths of the standard hulls, especially the later examples, their shiplike hulls remained difficult for nonship carpenters to build, and their loaded drafts were often too great for the sheltered waters where they were employed. A smaller, simpler to construct ironclad with a light draft was needed for many of the shallow waterways, particularly in the rivers and sounds of Virginia, North Carolina, and the west. Naval Constructor Porter prudently planned for such vessels beginning in the Civil War's early days.

SEVEN

EARLY ATTEMPTS AT AN ALTERNATIVE HULL FORM

The vessel is not so strong as I supposed her to be.
—Admiral Franklin Buchanan, Confederate States Navy

These two most obscure and by all accounts unusual Confederate ironclads are believed to be the first diamond hull vessels completed. The diamond hull is so called because in cross-section a vessel of this design presents a hexagonal shape, with no curves. Constructor John Porter intended these ironclads and others like them to be simple to construct, so their hulls contained no curvature laterally and only very little longitudinally.

Both the *Tuscaloosa* and the *Huntsville* were built under contract to the Navy Department by Selma, Alabama, shipbuilder Henry D. Bassett (who also built CSS *Tennessee*) beginning in May 1862. They are considered separately from other diamond hull designs because they apparently differed in construction characteristics. Little is known about their final configurations—no plans have been found, and only a few contemporary sketches are known to exist. Both were unsuccessful as self-propelled vessels. Constructor Porter is generally credited with the planning of these two ironclads, but they seem to have differed so much from his later diamond hulls that someone else may have designed them.[1]

Evidence indicates that the *Tuscaloosa* and the *Huntsville*, as well as two similar vessels laid down on Alabama's Tombigbee River but never completed, featured elements of western river steamboat construction, which did not appear in any of Porter's other designs. Most eyewitness accounts comment on the unusual hull of these ironclads; indeed, this appears to have been the biggest contributor to the *Tuscaloosa* and the *Huntsville*'s lack of motive power.

CSS *Tuscaloosa*

CSS *Tuscaloosa* was the first of the two completed, on February 7, 1863.[2] The ironclad was launched at Selma and taken to Mobile for fitting out. Very little firsthand information concerning the *Tuscaloosa*'s configura-

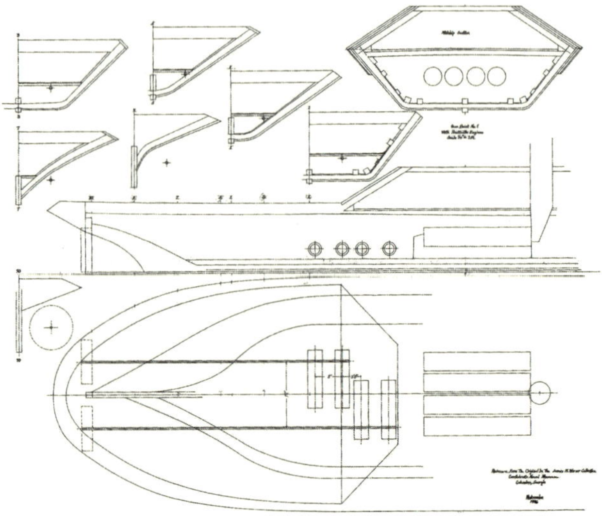

Figure 7.1. Partial plan view, inboard profile, and transverse sections of Gunboat No. 1 with Prattville Engines, redrawn by Bob Holcombe in 1976. Note the four western river–type boilers, distinguishable by their long cylindrical shape, narrow diameter, and positioning of the furnaces forward as revealed by the location of the single smokestack. Other unusual features are the four engines, two per propeller shaft, and the unusual chine (angular turn of the bilge) construction, reminiscent of western riverboats. The *Tuscaloosa* and the *Huntsville* are thought to have been similarly constructed.

tion has come to light. It was probably twin screw, like later diamond hull ironclads, but of unusual design "singularly shaped" and apparently not reflecting well on Confederate naval architecture according to contemporary sources. No plans are known, but a single draft labeled "Gun Boat No. 1 with Prattville Engines," preserved at the Port Columbus Civil War Naval Museum, appears to depict a never-completed sister ship, one of the Tombigbee River ironclads (fig. 7.1). Recent research has revealed that this vessel is probably not the *Tuscaloosa*, as originally thought; the unusual machinery and contemporary accounts provide the evidence.[3]

Measurements taken from the drawing show a diamond hull ironclad 43.5 feet in beam and 10.5 feet in depth of hold, with an estimated length between perpendiculars of 160 feet on a 9-foot draft.[4] These dimensions approximate those attributed to the *Tuscaloosa* in contemporary accounts.

The *Tuscaloosa*'s machinery source is better documented. As with the *Tennessee*, it was originally intended that the machinery for this vessel be built by Columbus Naval Iron Works, but there was not time or materials available to prepare purpose-built engines.[5] Instead, high-pressure riverboat machinery from the steamer *Chewala* was reworked for the *Tuscaloosa*.[6]

The modifications were made by William Penny & Company of Prattville, Alabama. This company was a branch of Columbus Naval Iron Works.

Not much is known about William Penny & Company, but it seems to have been an outgrowth of Penny & Chadwick of Milton, Florida. After the Union gained control of that area in 1862, William Penny apparently moved his works to Prattville, where it came under Confederate navy control.[7] This facility presumably had a fair level of capability to operate as a naval works auxiliary.

The manufacturer of the *Chewala*'s machinery has not been found. That vessel was a 372-ton sternwheeler built at Brownsville, Pennsylvania, in 1852. It served on the Apalachicola River before passing to Confederate control in 1861.[8] The boat's machinery was later sold to Columbus Naval Iron Works by that facility's former owner for two thousand dollars.[9] This included the *Chewala*'s boilers (number unknown). The machinery formerly driving the sternwheel was installed using a geared layout to drive the twin screws of the *Tuscaloosa*.[10]

The *Tuscaloosa*'s machinery configuration can be guessed: if the draft for Gun Boat No. 1 indeed represents a sister vessel, the *Tuscaloosa*'s machinery would have had similar dimensions, if not so complicated a layout. Gun Boat No. 1 is unusual for its four single-cylinder engines 30 inches in diameter and a 10.25-foot stroke turning twin propellers 7.75 feet in diameter. Steam was supplied by four boilers, each 3.25 feet in diameter and 19.5 feet long.

When the *Tuscaloosa* was finally completed at Mobile, its design remained to be tested. Admiral Franklin Buchanan supervised the machinery and speed trials of the vessel. His reports to Secretary Stephen Mallory are the only detailed firsthand testimony discovered so far that relate to the *Tuscaloosa*'s and later the *Huntsville*'s poor performance under steam.

In early April 1863 the *Tuscaloosa*'s trials commenced. It was quickly revealed that despite the seemingly suitable machinery installation and arrangement, the ironclad was far too slow. Buchanan reported the disappointing results to Mallory on April 6: "I have the honor to inform you that I made a second trial trip yesterday with the 'Tuscaloosa' and found the speed was not increased by using coal instead of wood, stopping the leaks in her boilers and increasing the draft by blowers [forced draft fans]. The machinery worked remarkably well, and she carried from 135 to 150 pounds of steam, 25 revolutions of the engine and only made about two and half (2½) knots per hour, her model is the cause of her want of speed. She scarcely stemmed the current in the harbor."[11]

This statement is revealing in several respects. First, it is important to note the attempt at using scarce coal in order to provide more heat and thus more steam for the engines than wood would provide. Second, the boiler leaks imply haste and a lack of quality control in the boilers'

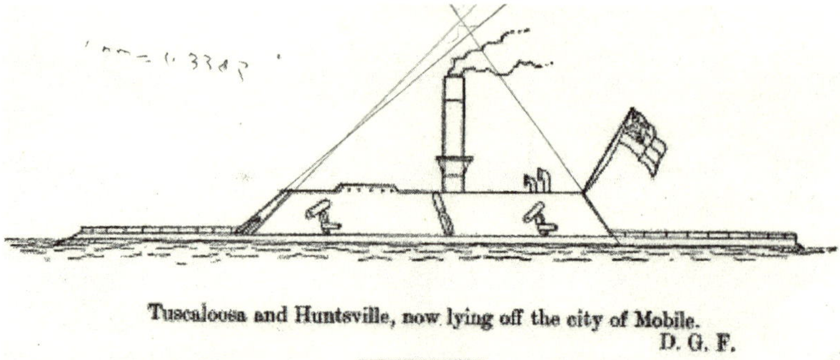

Figure 7.2. This sketch by Union Admiral David G. Farragut depicts the *Tuscaloosa* and the *Huntsville* as he saw them during the Mobile Bay campaign. Apparently the two appeared so alike (at least from a distance) that only one was worth depicting. The image agrees with the engineering descriptions that there were two ventilators. Also noteworthy is the design of the gunport shutters and pilothouse integral with the casemate; these strongly resemble CSS *Tennessee*, built by the same constructor, Henry D. Bassett.

installation. CSS *Arkansas* faced a similar problem, as related previously. Third, the installation of forced draft fans was a further, more expensive, attempt to improve both ventilation and speed. Only a few other ironclads were so fitted. Finally, Buchanan's assertion that the "machinery worked remarkably well" seems surprising considering all the problems, but marine machinery, no matter how powerful, still depends on a hull's hydrodynamic efficiency for effective propulsion. Buchanan's statement in this regard further corroborates criticisms of the *Tuscaloosa*'s hull design, although in exactly what way it was defective remains a mystery.

The *Tuscaloosa* was ultimately a failure as a self-propelled vessel. Consequently, it served out its existence as little more than a floating battery and was scuttled at the confluence of the Mobile and Spanish Rivers on April 12, 1865, by the retreating Confederates.[12] The wreck was located in 1983, but exploratory surveys of the seemingly well-preserved remains failed to uncover anything new.[13] Future in-depth investigation of this valuable Civil War archaeological resource has the potential to uncover a wealth of new data.

CSS *Huntsville*

CSS *Huntsville* was the sister ship of CSS *Tuscaloosa*, but even less is known about this second Selma ironclad. The *Huntsville* was built alongside the *Tuscaloosa* and launched the same day. The ship was less complete than the *Tuscaloosa* when launched and had to be towed to Mobile without its boilers and engines.[14] The origin and type of power plant eventually supplied by Columbus Naval Iron Works is unknown—all that is stated is that it was a high-pressure system taken from a riverboat.[15] This ironclad's engines, which like those on the *Tuscaloosa* could not move the vessel effectively, apparently were taken from a river steamer and reworked either at Columbus or Prattville.

The sidewheeler *John C. Calhoun* is a possible source of the *Huntsville*'s machinery. This vessel was built in 1859 at West Brownsville, Pennsylvania, and displaced 165 tons.[16] The *Calhoun* exploded, with a loss of human lives, near Bainbridge, Georgia, on April 28, 1862.[17] The machinery was salvaged and the Columbus works made a breech-loading cannon from one of the shafts. The engines were supposedly the same size as those from the *Chewala* installed on the *Tuscaloosa*.[18]

To what extent the completed *Huntsville* resembled its sister is uncertain, but ultimately it performed just as poorly (fig. 7.2). After a relatively short (for a Confederate ship) fitting-out process, the *Huntsville* began undergoing trials at Mobile in May 1863. Admiral Buchanan again supervised. His report to Secretary Mallory was basically the same as for the *Tuscaloosa*: "I have the honor to forward for your information copies of the reports of Lieut. Comdg. Julien Myers and 2nd Asst. Engineer C. M. Jordan relating to the trial trip of the steamer 'Huntsville' made on the 11th inst., during the trial I found the machinery to work well and the speed about 3½ knots. The heat in the fire room will not be so great when the fans are put in her and the ventilators cut which I have directed to be done. The vessel is not so strong as I supposed her to be, she trembled considerably."[19]

Once again, forced draft fans were installed in an effort to improve the speed, but it is the mention of ventilators that is interesting. Generally, ventilators seem to have been installed on ironclads mostly for crew comfort, instead of what appears to be the more obvious purpose of directing airflow into the machinery spaces. Buchanan at least seemed to have recognized this problem, but why it was not addressed in other cases is unknown.

A final point to be made about the *Huntsville* is the mention of the ship trembling while underway. This was probably a combination of poor construction, inefficient hull design (water resistance), and the heavy gearing used to connect the adapted riverboat engines to the screws.

The *Huntsville* served out the war alongside the *Tuscaloosa* as a floating battery, its apparently good machinery being no match for poor design. Again, what element(s) made it poor are currently unknown. Unlike its sister ship, the *Huntsville* did make a minor contribution to the final Confederate defense of Mobile. It aided CSS *Nashville* in bombarding the Union troops that attacked Spanish Fort and Fort Blakely in late March 1865.[20] Both ironclads and their escorts were severely limited in rate of fire by the lack of ammunition at that time, and the last defenses of Mobile fell on April 8. When the remaining Confederate naval forces retreated upriver shortly after, the *Tuscaloosa* and the *Huntsville* were together again, and together they were sunk by their own crews when their lack of steaming power became a crucial liability.[21] Their wrecks, apparently almost completely intact, lie close together near Mobile awaiting further investigation.

Conclusion

The *Tuscaloosa* and the *Huntsville* remain two of the most mysterious and tantalizing Confederate ironclads. The lack of concrete data on their design, machinery, and even size and armament, coupled with their apparent state of preservation, make them ideal targets for an archaeological investigation. For these reasons it is also difficult to fit these two unusual sisters into the general classification scheme of Confederate ironclad types. There seems to be enough information to safely assume, at least, that they were indeed diamond hull types with twin screws. Notably, unlike the later diamond hulls (such as CSS *Albemarle*), the *Tuscaloosa* and the *Huntsville* performed poorly despite their machinery. Admiral Buchanan attributed this to their hull design, an important and unique distinction among Confederate ironclads. Virtually every other ironclad suffered primarily from inadequate machinery, even though few of the hull forms used for Confederate ironclads were very hydrodynamic. Buchanan and his engineers either vastly overestimated the power output of the *Tuscaloosa*'s and the *Huntsville*'s geared riverboat engines, or their hulls were very poor indeed.

The crews on both these ironclads were forced to resign themselves to duty on ships that could function only as floating batteries. Living conditions on board were terrible even with improved ventilation; off-duty sailors actually had to sleep ashore in cotton warehouses at least part of the year.[22] Accounts of daily life on these ironclads are minimal, and any firsthand accounts of their steam plants in operation have not been found. Archaeological investigation of these two wrecks, more than of many other ironclads, stands to reveal details not only of the ships themselves but also of the daily life of their crews.

Ultimately, although the *Tuscaloosa* and the *Huntsville* were capable of some movement, it was not enough to allow them to participate in any important action or to escape destruction when Mobile finally fell. Their model was not repeated except in two of the unfinished Tombigbee River ironclads. The next group of diamond hull ironclads designed by Constructor Porter was the culmination of the type; one, at least, had a significant impact on the Confederate war effort in coastal North Carolina.

EIGHT

DIAMOND HULL IRONCLADS

The ironclad on the Roanoke . . . is considered . . . a formidable affair, though of light draft.
—Rear Admiral Samuel P. Lee, US Navy

A type of simplified ironclad hull suitable for shallow rivers and sounds, especially the inland waters of North Carolina, was designed by Constructor John Porter beginning in 1862. A relatively small size and an easily constructed hull with few curves characterized the four vessels that were completed: the *Albemarle*, the *Neuse*, the *Fredericksburg*, and the *Missouri*. Unlike the previous two diamond hulls (the *Tuscaloosa* and the *Huntsville*), these are well documented, and at least one made an important contribution to the naval war effort. All were twin screw except the *Missouri*, which was the only completed Confederate centerwheel ironclad; all except the *Fredericksburg* used secondhand machinery. For the most part, the machinery systems of these ironclads appear to have worked fairly well despite the boxy hull design.

Porter intended that the diamond hull vessels could be built quickly without skilled labor at protected inland sites and could operate in much shallower water than other standardized types like the *Richmond* class. As we will see, a mighty effort was made to build and equip the *Albemarle*, the *Neuse*, the *Fredericksburg*, and the *Missouri* in a timely manner at each of their scattered locations, but the lack of iron, unreliable transportation, and a shortage of labor once again caused significant delays. Construction times for these simplified designs were therefore not significantly shorter than those for other ironclads. Nor was Porter entirely successful in his calculations of draft: he planned a depth of 6.5 feet for the *Albemarle*, for example, which was exceeded by 1.5 feet.[1] All the others encountered similar underestimates.

Despite their problems, however, the diamond hulls were still a significant improvement in Confederate ironclad design, especially over the earlier *Tuscaloosa* and *Huntsville*. They were probably the first ironclads built with the full realization of what Southern industry was and was not capable of in rural areas, as well as of operating environment.

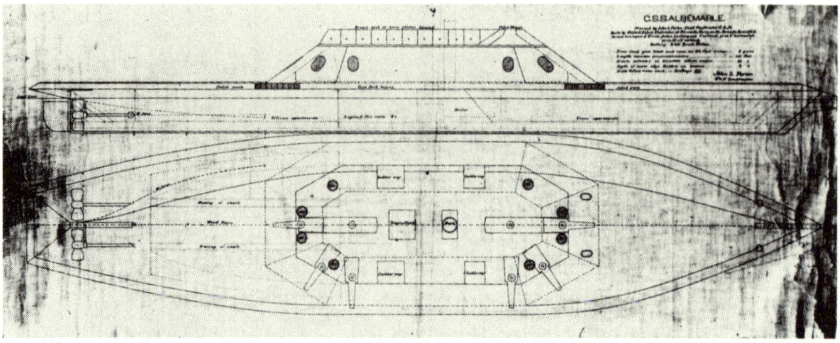

Figure 8.1. Original plans of the *Albemarle*. The ship as completed was very similar to the layout shown, except that the quarter panel gunports were not cut (they were kept for the sister ship *Neuse*), nor was the hinged iron bulwark for sharpshooters on the casemate top installed. Note the horizontal boiler with the aft-facing furnace shown amidships in the elevation view at the top. The engines are not shown.

CSS *Albemarle*

The shallow-draft diamond hull was well suited to the rivers and sounds of coastal North Carolina, and the most famous example, CSS *Albemarle*, was completed in that state. This vessel and its sisters differed from the *Tuscaloosa* and the *Huntsville* based on considerable surviving evidence. A simplified design, hexagonal in cross-section, with integrated casemate and shallow keel, was laid out by Constructor Porter (fig. 8.1). The *Albemarle* was the first of these ironclads to be completed and closely followed this pattern. The design was 139 feet long between perpendiculars, 155 feet long overall, 34 feet in beam, 9 feet in depth of hold, and 6.5 feet in draft.[2]

The *Albemarle* was laid down near the Roanoke River at Edward's Ferry, North Carolina, in early January 1862 under contract to Colonel William F. Martin and Lieutenant Gilbert Elliott.[3] Elliott was only nineteen years old when construction began, but by that time he had already made important contributions to the naval defense efforts of his home state.

Gilbert Elliott was born on December 10, 1843, in Elizabeth City, North Carolina, the son of a prominent local lawyer. Until war intervened, it seemed that young Elliott would follow his father's profession. Instead he became a junior naval contractor after North Carolina seceded through his friend and mentor Major James G. Martin, the brother of Elliott's later employer and business partner William Martin. Through these prominent connections, the young man served first in maintaining the Martins' shipyard, then with their influence graduated to becoming an independent naval contractor. By 1863 Elliott was promoted to army lieutenant, personally met Secretary Mallory, and was awarded several contracts: to build gunboats at Elizabeth City, North Carolina, ironclads at Norfolk (two of the Dismal Swamp ironclads), and an ironclad at Tarboro, North Carolina. Unfortunately for the young shipbuilder, none of these projects came to fruition; first the Albemarle Sound area, then Norfolk, came under federal control. Elliott's base of operations was moved to the inland site of Halifax, North Carolina, on the Roanoke River by the time he was awarded the contract for the *Albemarle*. Since there was not enough room on the riverbank at Halifax for construction of an ironclad, the nearby but more secluded site of Edward's Ferry was chosen. The vessel that became Elliott's crowning achievement was soon under construction there.[4]

Throughout his wartime career Elliott was fortunate to have the friendship and influence of several prominent men. One of these was the *Albemarle*'s first captain. Commander James W. Cooke was also a North Carolinian, born in Beaufort in 1812. He joined the US Navy as a midshipman at age sixteen and by 1841 was promoted to lieutenant. He continued in that post until resigning his commission on May 2, 1861, but he soon received the same appointment in the Virginia state navy. Cooke then became a lieutenant in the newly formed Confederate States Navy in June 1861. As a naval officer, he became the official liaison between contractors, such as Martin and Elliott, and the government. He was responsible for all such projects in North Carolina by the time the *Albemarle* was begun, and he became closely involved in the construction effort, eventually being assigned as the superintendent. Cooke's enormous organizational and scrounging abilities were immensely helpful to Elliott during the *Albemarle*'s building. He rounded up men, materials (most important, scarce railroad iron), and support. As a result, Cooke was offered the command of the *Albemarle* in January 1864. He gratefully accepted the well-deserved promotion.[5]

Through Elliott's and Cooke's efforts, the *Albemarle* was launched incomplete after fifteen months' construction time, on October 6, 1863. Some damage to the keel was incurred, but the small ironclad was towed upstream to Halifax and completed there in April 1864. The finishing touches were still being put on even as the *Albemarle* was commissioned on April 17 and steamed downriver to attack federal forces at Plymouth. The ironclad's final design differed from the original plans, with an overall length of 158 feet, an extreme beam of 35 feet 3 inches, a width on deck of 32 feet, a depth of hold of 8 feet 2 inches, and a loaded draft of 9 feet.[6] These dimensions come from a May 18, 1864, Union survey of the ship after it was raised, taken to Norfolk, and stripped of its armor. Other sources suggest that the length was decreased slightly, to 152 or 153 feet. The four corner gunports were also not cut, leaving only six in total.[7]

During the *Albemarle*'s construction, one of Cooke's greatest headaches was the procurement of propulsion machinery. Where it was obtained remains something of a mystery. Several conflicting accounts exist. The commander of the *Albemarle*'s sister ship the *Neuse*, Benjamin F. Loyall, remarked years later that much like his own vessel, "the engine of a large saw mill was altered and made to serve for her [the *Albemarle*'s] propelling power."[8] Conversely, a publication around that same time mentioned that the two engines were from Tredegar Iron Works in Richmond, each turning a propeller and with 100 nominal HP.[9]

In addition, former Confederate officer (and author of the first complete history of the Confederate navy) J. Thomas Scharf included a quote from one of the *Albemarle*'s former commanders, John N. Maffitt: "The engine was adapted from incongruous material, ingeniously dovetailed and put together with a determined will."[10] It was also stated that only "a simple blacksmith-shop aided the mechanical part of her construction."[11] It seems unlikely that the *Albemarle*'s machinery was built totally from scratch because the builders had difficulty finding any local iron, even simple railroad rails, for armor. The sawmill hypothesis is reasonable, but the mention of Tredegar throws all into doubt. Either way, Tredegar certainly had the ability to produce the *Albemarle*'s engines and is known to have provided the armor for the vessel. In addition, it is known that the Charlotte Navy Yard forged the propellers and shafts.[12] The provenance of this famous ironclad's boilers and engines will probably remain uncertain, since the *Albemarle* was broken up in 1867.

Whatever its source, the machinery stood up well in its early baptism by fire. Cooke and his new command were urgently needed for a renewed Confederate offensive in eastern North Carolina in April 1864. The *Albemarle* had to be navigated down the Roanoke to the strategically important town of Plymouth in order to support a Confederate attempt to retake it. Control of Plymouth would give the Confederates an important base from which to operate on and potentially regain Albemarle and Pamlico Sounds. The value of using armored vessels in the confined waters of North Carolina was finally going to be demonstrated.[13]

The last touches were still being put in place on the *Albemarle* as Commander Cooke piloted the ship downstream stern first. This unique practice was done to help counteract the river's current with its many twists and turns, making it easier to steer the vessel. En route there were some minor troubles with the shafting and the rudder, but this was not unexpected for a new ship that was steering stern first. All problems were overcome, and Cooke continued, drilling his crew all the while. By the time the ship approached Plymouth, both Union and Confederate forces in the area were preparing for the battle expected within the next few days. The local Union naval force consisted of the gunboats USS *Southfield* and USS *Miami*, with their escorts. The Union knew of the *Albemarle*'s approach but expected to easily "whip her," in the words of Commander Charles Flusser, who was in charge of the US naval force. On the evening of April 18, 1864, Cooke anchored the *Albemarle* about 3 miles above Plymouth, preparing for the battle to come.[14]

In the early morning hours of April 19, the ironclad cleared the submerged obstructions and shoreline batteries of the Union forces around Plymouth and proceeded to attack the Union ships coming upriver. USS *Southfield* and USS *Miami* were dragging heavy chains between them in an attempt to capture the *Albemarle*, but Commander Cooke easily escaped the trap and rammed the *Southfield*, sinking it in minutes. The ensuing melee saw the *Albemarle* almost dragged under by the *Southfield*, but the ironclad broke free and attacked the *Miami*, killing Commander Flusser when a shot he himself fired rebounded off the *Albemarle*'s casemate and exploded at his position. Demoralized, and after sustaining heavy damage (apparently inflicting none on their opponent), the Union forces retired downriver. Only one man was killed on the *Albemarle*, and the ship had virtually no damage.[15] The Union forces in Plymouth took a beating that night: with their waterborne support and communications cut off, they were forced to surrender by the following day.[16] Meanwhile the little ironclad from a riverside cornfield steamed to Plymouth in triumph. The *Albemarle* had achieved an important strategic victory against overwhelming odds. Cooke and his ship joined the narrow ranks of the Confederate ironclads and their captains that acquitted themselves well in battle.

For the next two weeks the *Albemarle* remained at Plymouth, which again became a major Confederate regional base. The ironclad posed a constant threat to local Union forces, and the Confederates confidently set out to reconquer many areas under federal control. They were largely successful because of the threat of using the *Albemarle*. On May 5 the ironclad was sent out to support an attack on New Bern (North Carolina). This meant a journey into the open waters of the sounds. In company with the lightly armed gunboats *Bombshell* and *Cotton Plant*, the *Albemarle* encountered a fleet of seven Union warships under the command of Captain Melancton Smith. The US vessels were the *Mattabesett*, the *Sassacus*, the *Wyalusing*, the *Whitehead*, the *Ceres*, the *Commodore Hull*, and the *Miami*.[17] The Confederate ships soon found themselves involved in a closer action than that of April 19. Intense fire was thrown up on both sides, and USS *Sassacus* zealously attempted to sink the *Albemarle* by ramming, sustaining heavy damage in the process. Although little apparent damage was done to the Confederate ship's main structure, the Union slowly whittled away its steaming power by destroying the smokestack.

Finally, the engagement ended inconclusively with the *Albemarle* retreating from the badly damaged Union vessels. Even though the exterior of the ship received minor damage, the lack of a functioning smokestack meant that only when the crew threw butter, lard, and bacon into the boilers could they raise enough steam for retreat.[18] Arriving safely back at Plymouth, the *Albemarle*, after repairs, was still able to claim ascendancy on the rivers and sounds of North Carolina. During all encounters its steam machinery apparently functioned quite well. The only obvious flaws that can be observed without any firsthand data from the engine room were those typical of most other ironclads: slow speed and vulnerability of the smokestack to battle damage. As Cooke experienced in the May 5 battle, lack of forced draft for the boiler fires could become a critical weakness if the stack was severely damaged.

By June 1864 Commander Cooke had been suffering from ill health for some time. He requested Secretary Mallory to relieve him of his command so he could recuperate, and the request was granted. For his excellent achievements in battle, Cooke was awarded the rank of captain on his departure. On June 9 Commander John N. Maffitt was appointed the *Albemarle*'s new master. He had little to do, since his predecessor had established Confederate supremacy in the region. Maffitt had gained fame for his service on the commerce raider CSS *Florida*; he probably chafed under the restraint and inactivity of his new command. The *Albemarle* was not allowed to venture out; as a one-ship "fleet in being," the ironclad was far too valuable to risk losing in another open water battle against superior numbers. Therefore, for the rest of the summer of 1864 and into the early fall, the *Albemarle* remained a mostly stationary deterrent to Union forces contemplating any attacks in coastal North Carolina.[19]

Moored safely behind a floating log boom at Plymouth (for protection from ramming by any enemy vessels that might approach and, more significantly, spar torpedoes), the ship represented the one obstacle keeping the Union forces at bay in the region. The *Albemarle*'s mere presence was thought to be able to deter any future Union incursions, and for a time this was true. No one among the Confederates could have foreseen the circumstances that led US Navy Lieutenant William B. Cushing and a spar torpedo to send the *Albemarle* to the bottom. Despite all evidence to the contrary, by September 1864 the "invincible" *Albemarle* only had one month to live.

On September 9 Commander Maffitt was transferred to command the blockade-runner *Owl*, and the *Albemarle*'s next and last captain received orders to report to Plymouth. That man was Lieutenant Alexander F. Warley, already a veteran ironclad captain and participant in the naval theaters of New Orleans and Charleston.[20] Warley was in command of the steamer *Water Witch* at Savannah when he received the assignment. After making the long journey to eastern North Carolina, he was able to size up his new ship and separate the myth (the *Albemarle*'s exploits were already famous all over the country) from the reality before him. The ship was moored to the town wharf with the bow downriver, repaired from the last battle, but it was certainly not the invincible giant the Union often believed it to be. The crew was composed of about sixty men, most of whom were landsmen used to sleeping ashore, and many were sick.

Nevertheless, Warley, in his usual bold way, seized immediate control of the situation and took the *Albemarle* out to put on a show of force. There was little else he could do, however. The ship was basically the only thing keeping Plymouth in Confederate hands, and it could not be risked by steaming into open water and potentially losing the town. Warley, like Maffitt before him, did as much as he could in such a cramped situation, but for a little more than a month he and his ironclad remained mostly static. Meanwhile, reconnaissance parties of Union sailors continually scouted the swamp on the far side of the river, looking for a chance to put their plans of capturing or destroying the *Albemarle* into action. These plans had been underway for some time.[21]

Union forces in North Carolina were very intent on removing the threat posed by the little ironclad. By October 1864 virtually nothing had been done to solve the problem, but a plan forwarded in midsummer by Lieutenant Cushing, a daring young officer, was eventually decided upon. It was just unusual and desperate enough to work: using a steam-powered launch, Cushing and about fourteen men would attack the *Albemarle* at its moorings and attempt to sink it using a torpedo attached to a spar on the launch's bow.[22] Finding volunteers for the dangerous mission was not difficult because Union sailors were eager to take revenge for their humiliations during the summer. Therefore, on the night of October 27–28, 1864, Cushing and his men moved on Plymouth. The lightly constructed launch was able to elude the pickets and slide over the protective but slippery log boom surrounding the *Albemarle*. Cushing himself pulled the trigger of the torpedo as it rammed into the ironclad's hull at the waterline. Amid the resulting explosion and a volley of rifle fire from the Confederates, only Cushing and one other man escaped; all the others were killed or taken prisoner. Despite the casualties and the odds, however, they had succeeded.[23]

Lieutenant Warley was on the gun deck inside the casemate when the fatal blow was dealt. Amid the general confusion of small-arms fire, and immediately after the deafening report of the *Albemarle*'s 6.4-inch bow gun, Warley and the others felt "a report as of an unshotted gun, and a piece of wood fell at my feet." The carpenter was sent below and soon reported a "hole in her bottom big enough to drive a wagon in."[24] The *Albemarle* rapidly settled to the bottom of the Roanoke.

With his ship out of commission and neither main gun trainable, Warley did what he could to prevent the Union from retaking Plymouth and raising his ship. Lying in shallow water, most of the superstructure remained dry and could be readily accessed. The crew salvaged what it could, but the ship was abandoned as resurgent Union forces captured Plymouth on October 31. There was now nothing left to stop them. In a last-ditch effort to prevent the gallant *Albemarle* from falling into Union hands, Warley had the crew place explosives to blow out one side of the casemate before they set out for Halifax.[25] The damage ultimately was not enough to render the vessel worthless. The Union refloated the ironclad, and it became an unofficial member of the US Navy. After languishing for some months in the Plymouth area, the *Albemarle* was towed to the Norfolk Navy Yard in April 1865.[26] The partly dismembered hulk became a curiosity there and was studied by the authorities for two years before being sold for twenty-five hundred dollars in October 1867 for breaking up.[27]

While the *Albemarle* was at Norfolk, a survey was made of it. Nearly all the available data on the ship's machinery comes from this report. It describes the *Albemarle*'s two engines in detail as the noncondensing horizontal type, 18 inches in cylinder diameter, with a 19-inch stroke and geared to twin three-bladed propellers 6 feet in diameter by four gear wheels. These wheels measured 2 feet 9.5 inches in diameter at the pitch line, and the pinions measured 2.5 feet in diameter. Powering the engines were two vertical watertube boilers with one furnace each, 15 feet 4 inches in length, 5 feet 2 inches in height, and 4 feet 7 inches in diameter. The furnaces each measured 4 feet 6 inches in length and 4 feet in width, and the flues were 1 foot 9 inches high and 9 inches in diameter. Each boiler was equipped with 120 watertubes and a steam drum 2 feet 8 inches in diameter and 1 foot 6 inches high. Also noted was that all pipes and steam lines were made of cast iron. Despite the ordeal the hulk had gone through by that time, the report nevertheless states that "the engines and boilers are in fair condition and need but few repairs."[28] It is unknown whether the *Albemarle*'s machinery was reused after the vessel's sale.

The *Albemarle* became one of the most famous and successful Confederate ironclads, and it justified Constructor Porter's design notions of diamond hull ironclads for shallow-water operations. It was also notably the only Confederate ironclad sunk by enemy action. Its sister diamond hulls had no such distinction.

CSS *Neuse*

The next diamond hull ironclad completed in North Carolina did not have nearly as much success or fame as the *Albemarle*, although the two can be considered sister ships. The *Neuse* was very similar to the *Albemarle*, but it conformed more closely to the original plans.[29] The ship was laid down at Whitehall (now Seven Springs) on the Neuse River in October 1862 and completed by contractors Thomas S. Howard and Elijah W. Ellis. During construction, concerns over the ship's draft led to a significant reduction in armor plate. Changes were otherwise minimal, and the *Neuse* was launched in mid-March 1863. Its completion was delayed by materials shortages after it had been towed downstream to Kinston. The extremely makeshift nature of its steam machinery reflected many such difficulties.[30]

Throughout its construction and service life, the *Neuse* seemed dogged by bad luck. The ship was nearly destroyed before it was even launched. Because the Whitehall site was a fairly isolated area, from the start a lack of building materials such as iron and mechanical parts hampered the construction process. Work at the site was slow and sporadic because of this, but there was another, more serious problem: repeated attacks by Union raiders. While still sitting incomplete beside the river, the *Neuse* was attacked and shelled by Union forces on December 16, 1862.[31] The unfinished, unarmored hull was not severely damaged by the raid, and the Confederates soon had control of the area again, but as repairs began, it became clear that the future ironclad should be moved in order to have any chance of completion.

When the *Neuse* was finally towed downriver to Kinston, finishing the project proved to be just as difficult as before, despite the more accessible location. For the better part of a year the ship awaited sufficient armor, armament, and machinery. As its sister the *Albemarle* was hurriedly completed at Edward's Ferry and driven into battle at Plymouth, the future CSS *Neuse* was slowly built up, fitted out, and furnished. By April 1864, with Union forces closing in from all sides, it looked as though the ironclad, after two years of building, would be finished just in time.

As in many other cases, the greatest delay in the *Neuse*'s completion was obtaining steam machinery. Like the *Albemarle*, the ship was built in an isolated area with no readily available engines. Therefore, the machinery for the *Neuse* had to be scrounged from any local source and adapted to drive the heavy ironclad. Records on how this was done were scarce and mixed with local lore, and for many years there was a good deal of uncertainty. It was long believed that the single 15-foot boiler powering the ironclad was taken from an 1844-built steam locomotive owned by the Baltimore & Ohio Railroad: no. 34, the *Gladiator*. The exact provenance of the engines likewise remained uncertain because several sources were listed. Among these are Pugh's Mill (presumably a sawmill) located in New Bern, North Carolina.[32] Other possible sources included Confederate Naval Works in Richmond and the cylinders of a steam locomotive, probably the same unit given as the source of the boiler.[33]

In 2013, researchers at the newly opened CSS *Neuse* Center in Kinston discovered definitive information revealing the actual provenance of the *Neuse*'s machinery. A look at the Baltimore & Ohio locomotive roster for the Civil War years reveals that the long-held belief in locomotive no. 34 providing the *Neuse*'s boiler is erroneous, because "the boiler of engine No. 34 was put in the gunboat *Flora McDonald* and was sunk in the river opposite this place [Wilmington, NC]." The boiler was removed from the locomotive sometime after its capture near Shenandoah Junction, Virginia by Stonewall Jackson's troops in early 1861.[34]

Pugh's Mill in New Bern was likewise revealed as the correct source of the *Neuse*'s two engine cylinders through a detailed look at Union Major General John M. Peck's extensive 1864 report on the state of affairs in eastern North Carolina: "The engine has been taken from Pugh's Mills, and was once in a factory in this city [New Bern]."[35] Although the dimensions of the machinery are not given, they probably closely approximated those of the *Albemarle*.

When finally completed, the *Neuse*, like the *Albemarle*, was desperately needed as backup for local Confederate forces, which by late April 1864 were planning an attack on the Union-held port of New Bern at the mouth of the Neuse River. The ironclad's two powerful 6.4-inch Brooke rifles would be an invaluable asset to the assault. Thus, under Commander Benjamin Loyall, the *Neuse* departed Kinston on April 27, heading downriver.[36] Low water levels caused constant worry over grounding. Sure enough, after steaming only about 0.5 mile, a crunching sound was heard and the ironclad ground to a halt on a shallow sandbar. All day long the crew tried desperately to work the *Neuse* off the bar as the river continued to fall, but by nightfall it was apparent the ship was totally stuck, since the bow was 4 feet out of the water. Five days later it had to be accepted that the *Neuse* would be forced to wait until the next large rainstorm to have any hope of floating again, and the attack on New Bern was canceled.[37]

The *Neuse* was rescued from the sandbar sometime in May 1864 and was soon back at Kinston.[38] Once moored, though, it was not able to venture out again. Trapped by the strongly defended obstructions at the mouth of the Neuse River and by the unavailability of troops to cooperate with it, the ironclad waited helplessly while Confederate forces, both land and sea, were systematically destroyed in coastal North Carolina. By the time Wilmington, the last port remaining in Southern hands, fell on February 22, 1865, the *Neuse* was the only remaining Confederate warship in North Carolina.[39]

By March 1865 the fighting reached the Kinston area. On March 9 the local Confederate commander, General Braxton Bragg, began to evacuate the town. The *Neuse* was detailed to cover the army's retreat and "if practicable, before sacrificing, [it was] to move down the river by way of diversion, and make the loss . . . as costly to the enemy as possible."[40] This order effectively resulted in a suicide mission, for the coal and provisions were enough for only a one-way trip. Traveling downriver very slowly (the maximum speed was only about 4 knots), the *Neuse* began shelling Union cavalry forces along the bank. With nowhere to go, the crew, presumably satisfied with the extent of the shooting, set its ironclad on fire and abandoned it. Fortunately for posterity, before the fires could reach the *Neuse*'s magazine, the intense heat ignited a loaded cannon, blowing a gaping hole in the ship's bottom. Muddy water rushed in, and as the little warship settled into the deep mud of its namesake river, the fighting in the coastal regions of North Carolina largely came to an end.[41]

The *Neuse*'s machinery was never able to move the vessel very effectively, but there are no reports of it being unreliable or needing repairs. This was probably at least in part because of general inactivity. In any case, after the hostilities ended, the ironclad's valuable materials, especially the iron armor and machinery, were not forgotten. The machinery seems to have been removed by New Bern treasury representative William Heaton. He was able to auction the *Neuse*'s machinery to the New York firm of Satterlee & Lyon for thirty-five hundred dollars.[42] The auction advertisement in the October 3, 1865, edition of the *North Carolina Daily Times* listed two engines, one donkey engine, and one large boiler "of eighty horsepower . . . by order of the Supervising Special Agent E. Hubbs."[43] Unfortunately, the type of boiler is not stated.

Another story concerning the removal of machinery from the wreck states that it was taken by Captain Joseph M. White for use in his Kinston sawmill. Supposedly White received government permission for the salvage and later sold the machinery to O'Berry Manufacturing Company in Kinston.[44] No documentation of this turn of events has yet been found.

The wreck of the *Neuse* was excavated and raised from 1961 to 1966. It is one of only four Civil War ironclads so preserved and is on display in Kinston. Unfortunately, the salvage effort resulted in the destruction of large portions of the wreck, but enough of the lower hull remains for valuable construction information to be learned. Archaeological evidence in the form of wooden mounting blocks for the engines and boiler shows that the engines were mounted next to each other on the starboard side to balance the weight of the boiler to port and must have been geared to the twin propeller shafts. The gearing arrangement was probably similar to that on the *Albemarle*.

CSS *Fredericksburg*

Experience with the *Albemarle* in battle showed that despite the design's strengths, the two-gun armament resulted in a slow rate of fire and a small broadside for the size of the ship. In fact, these issues were Commander James Cooke's primary criticisms when he reported his victories at Plymouth to the Navy Department.[45] Constructor Porter actually seems to have anticipated the firepower issue. At about the same time that the *Albemarle* and the *Neuse* were laid down in North Carolina, a larger four-gun diamond hull ironclad was begun at Richmond.[46]

The *Fredericksburg* was the first ironclad built entirely in Richmond, at a newly established shipyard in the Rocketts suburb.[47] The vessel was laid down in the spring of 1862 and in virtually all respects represented a stretched version of the *Albemarle* and the *Neuse*. As a result, the *Fredericksburg*'s dimensions were increased to 170 feet long between perpendiculars, 188 feet overall, 40 feet in beam, 10 feet 10 inches in depth of hold, and 9 or 10 feet in draft.[48] In overall appearance it closely resembled the *Albemarle*, but it had twin pilothouses and a rectangular casemate.

Although the *Fredericksburg* was built in Richmond at the center of the Confederate naval industry, remarkably few details of its machinery have been found. Its exact source is uncertain. As previously stated, Tredegar Iron Works contracted to supply the machinery for the ironclads built in Richmond and may have built the engines for the *Fredericksburg*. Conversely, the contract called only for supplying engines, so construction may very well have been outsourced, particularly to Confederate Naval Works (the Shockoe Foundry). This is the current thought on the installation of the ship's boilers and other mechanical components.[49] No definitive data have yet been found.

The *Fredericksburg* was successfully launched on June 11, 1863, after a failed attempt on June 6, but it was not completed until early 1864 because of the usual lack of materials.[50] Much of what is known of the ship comes from the extant plans prepared by Constructor Porter (fig. 8.2). Provided no changes were made after the draft date on the plans, the *Fredericksburg* was apparently powered by three cylindrical boilers approximately 15 feet long, 5 feet in diameter, and 7 feet high. The engine layout is shown by a single circle scaled to 24 inches in diameter. There is no way of knowing whether this represents the actual engine size as installed, because no other information has been found. Two propeller shafts are shown turning twin screws seven 7 in diameter, but the method of their connecting with the engines is also not shown—they may have been geared like the *Albemarle* and the *Neuse*.

The *Fredericksburg* joined the James River Squadron upon comple-

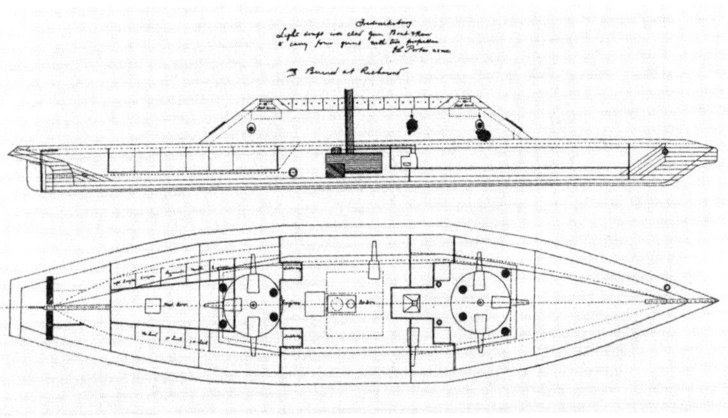

Figure 8.2. Plans of the *Fredericksburg*. The writing at the top says, "Light draft iron clad gun Boat & Ram, to carry four guns with two propellers—J. L. Porter, C.S.N.C. Burned at Richmond." Note the hinged armored bulwark for sharpshooters on the spar deck; it is thought not to have been installed. The interior layout is well detailed except for the machinery installation. The boilers are clearly shaded in, with their furnaces facing aft, and even the galley stove and its exhaust connection to the main smokestack is shown. The engines are unfortunately represented only by a vague circle in the profile view at the top. The twin propellers can be seen in the plan view at the bottom.

tion, but like the other vessels it remained effectively bottled up for most of its career. The only engagement the ironclad took part in was also the only battle fought by the James River Squadron: the Battle of Trent's Reach in January 1865. During that fiasco the Confederates tried to force their way past the river obstructions to attack the federal supply base at City Point. The *Fredericksburg*'s light draft meant that it was the only ironclad able to pass the obstructions. Unfortunately, the ship's lighter construction (it was the weakest ironclad in the squadron) put the *Fredericksburg* at great danger from Union shells, and the hull incurred damage from running across a sunken hulk. It was fortunate that the ironclad did not encounter the powerful 15-inch guns of the monitor USS *Onondaga* while awaiting the other Confederate vessels. In the end the waiting was in vain, and the squadron was forced to retreat upriver (the *Onondaga* finally entered the fray only as the Confederates retreated). Back at the squadron's home base near Drewry's Bluff, the *Fredericksburg* was repaired but did not venture forth again. As the sole ironclad capable of clearing the obstructions, it may have been able to inflict damage on the Union in a suicide run downriver, but it was never given the chance. In the final months of the war, the last Confederate fleet awaited the end, existing only to defend the capital at Richmond.[51]

The *Fredericksburg* never had the chance to prove its design as the *Albemarle* (and even the *Neuse*, to an extent) did. When Richmond was abandoned, the *Fredericksburg* was blown up and sunk to prevent capture

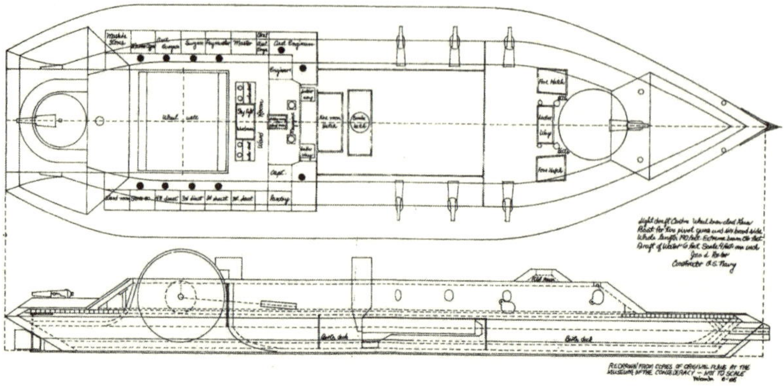

Figure 8.3. Plans of John L. Porter's light draft diamond hull centerwheel ironclad, redrawn by Bob Holcombe. The only ironclad completed with this general layout was the *Missouri*, although it was altered to carry two pivot guns forward; one pair of broadside guns and the exposed stern gun position were eliminated. Note the long western river flue boilers positioned with the furnaces facing aft and the inclined engine (represented here by a rectangle) connected to the center paddlewheel.

along with the other James River Squadron vessels on the night of April 2–3, 1865. Its remains still lie in the mud of the river bottom. Although the wreckage appears to have been heavily disturbed over the years, there may be enough left to allow a better understanding of this little-known diamond hull ironclad through archaeological investigation.

CSS *Missouri*

In addition to the light draft twin-screw diamond hull ironclads, Constructor Porter designed a vessel of similar form and draft driven by a center (or recessed) paddlewheel.[52] The first and only example of this ironclad type constructed as designed was CSS *Missouri*, laid down in December 1862 and completed by steamboat captains Thomas Moore and John Smoker.[53] This vessel was built and successfully completed within a relatively short time at Shreveport, Louisiana, on the Red River despite a lack of facilities, materials, and labor in the trans-Mississippi region.[54] Construction on the western frontier was quite a challenge, but the *Missouri* was furnished with effective armor and propulsion, as later reported by Union surveyors.[55]

The *Missouri* was launched on April 4, 1863, and retained the same layout on different dimensions from the original plans, which illustrate a vessel 193 feet long overall, 56 feet in molded beam, 58 feet in extreme beam, 9 feet 9 inches in depth of hold, and 6 feet in loaded draft (fig. 8.3). The *Mis-*

Figure 8.4. Watercolor by Union Ensign David M. Stauffer. This shows what can be considered an accurate view of the *Missouri*'s appearance at its surrender. Stauffer painted several vessels during his service with the Mississippi Squadron, and his paintings compare favorably with contemporary photographs. Note the steam capstan and the raised boarding ramp on the bow. The unusual ram has not been verified by any other source.

souri as completed measured 183 feet in overall length, 53 feet 8 inches in beam, 10 feet 3 inches in depth of hold, and 8.5 feet in draft (fig. 8.4).[56]

The *Missouri* shared a number of characteristics with a widely used class of Union river ironclad: the seven City class vessels constructed in the autumn of 1861. These were also centerwheel casemate ironclads with similar dimensions. They measured 175 feet long on deck, 51 feet 2 inches in beam, 6 feet in depth of hold, and 6 feet in draft.[57] One of these vessels, USS *Cairo*, was sunk by a torpedo (mine) on December 12, 1862, and raised 102 years later, in December 1964.[58] The *Cairo* is now on display at Vicksburg, Mississippi, and represents one of four preserved Civil War ironclads. It is also the only one with paddlewheel machinery, making for an effective comparison with the Confederate *Missouri*. It is unknown whether the design of these two ironclads was conceived independently for operation in a similar environment or knowingly adapted for Confederate use by Constructor Porter.[59]

Although the *Missouri*'s design is fairly well documented, there is some confusion about the source of its machinery. There appear at least two good candidate boats operating locally during the *Missouri*'s construction, and there are several hints in the primary sources.

The first candidate riverboat is the *Grand Era*, originally built at Louisville, Kentucky, as the *R. W. McRea* in 1853. The vessel was rebuilt and renamed the *Grand Era* in December 1860 after its sale to the Red River Packet Company. It was a sidewheeler measuring 240 feet long, 40 feet in beam, and 7.5 feet in draft.[60] Specific machinery details have not been found.

The steamboat *T. W. Roberts* is another possible candidate.[61] This vessel was a 288-ton sidewheel packet measuring 168 feet long and 32 feet wide with a depth of hold of 6 feet 8 inches, built at New Albany, Indiana, in 1860.[62] Its recorded machinery specifications differ slightly from the *Missouri*'s, but they are close enough to warrant a further look: an engine cylinder diameter of 20 inches with a stroke of 6.5 feet, powered by three double-flue boilers, each 40 inches in diameter and 28 feet long. All mechanical components were furnished by the ubiquitous Lent, South, & Shipman of New Albany, Indiana. The *T. W. Roberts* is noted as having a steam capstan, at that time fairly uncommon on river steamers; perhaps not coincidentally the *Missouri* also carried one.[63]

Unfortunately, there are contradictory accounts regarding these two boats, both of which were operating around Shreveport at the time of the *Missouri*'s construction.[64] Even though at least one source states that the *Missouri*'s engines came from the *Grand Era*, it is also suggested that the *Grand Era* was operating from Shreveport to New Orleans after the *Missouri*'s completion in late 1863.[65] In addition, another source states (mistakenly?) that the *T. W. Roberts* was renamed the *New Era* based on correspondence suggesting that the *New Era* be used to tow the *Missouri* downriver for completion. Further contemporary correspondence shows that the boat purchased for the *Missouri*'s machinery was owned by Russell B. Roberts.[66] The only steamboat that may fit the time and place for this is the *T. W. Roberts*. This evidence comes from statements in contemporary correspondence noting the former owner of the boat that was purchased for the *Missouri*'s machinery.[67]

It is certain that in 1863, the naval officer in charge of the *Missouri*'s construction purchased a boat's engines and boilers for sixty-five thousand dollars, and an engineer was sent by the Navy Department to refurbish them. The actual provenance of the power plant will remain unknown unless more conclusive information is found on the history and machinery of the *Grand Era* and the *T. W. Roberts*.[68]

Conversely, the *Missouri* itself is one of the best-documented Confederate ironclads in terms of its construction and career. Extensive research by Katherine B. Jeter has resulted in two important documents that collate a number of primary sources on the *Missouri*. The most important of these for the purposes of this work are the letters of the ironclad's devoted commander, Lieutenant Jonathan H. Carter. Carter's notes, which span the majority of the *Missouri*'s largely uneventful career, reveal much about operating a centerwheel ironclad on the Red River in Louisiana toward the Civil War's end.[69]

Carter was a North Carolinian who resigned his US Navy commission (ranked lieutenant) in April 1861. His pre-1863 Confederate service included duty on the gunboat *General Polk* at New Orleans and at the Jackson, Mississippi, naval station. Carter supervised the *Missouri*'s construction and the Red River defenses, finishing his wartime career as the captain of the ironclad. By all accounts he was a well-organized and courageous leader; his achievement in getting the *Missouri* completed is noteworthy.[70] Fortunately his detailed notes and letters are preserved at the Naval History and Heritage Command in Washington, DC.

Overseeing the gathering of materials and construction of the *Missouri* on the trans-Mississippi frontier was a challenging task for Carter, and he was especially anxious about the installation of his ironclad's machinery. He wrote to Secretary Mallory on February 2, 1863, "The contractors have purchased a steamboat [possibly the *T. W. Roberts*] with suitable machinery. The price paid was sixty five thousand dollars." He continued on April 4, "The machinery and boilers are on board and will be put in position immediately."[71]

Apparently there was a problem that caused a significant delay in installation, for on April 21 Carter wrote, "I am very anxious to get her machinery in position before she leaves here, as a steamboat to aid her will then be unnecessary. The [iron] cladding is being put on as rapidly as possible, and within three or four weeks I hope to have her machinery in position." He continued to report to Mallory fairly regularly, writing on April 29: "The boilers are in position, and the remainder of the machinery will be put in place as rapidly as possible." This was finally done, and by June 1863 construction was complete enough for Carter to begin to appraise the *Missouri*'s design. On June 15 he wrote, "As I have stated in a former letter, I think the centre wheel plan will prove slow, notwithstanding the power is more than sufficient to turn the wheel." The ship was apparently ready for trials by that time.[72]

On June 20, 1863, the first of many letters from Carter to various naval leaders regarding his new ironclad testified on the *Missouri*'s operational characteristics: "On the 17th inst., the 'Missouri' made a trial trip; the rate of speed obtained was six miles per hour upstream." On June 21 he wrote a detailed order to two engineers participating in the *Missouri*'s completion:

> You will answer the following interrogatories in regard to the engines, boilers, and machinery of the C.S. Str. "Missouri," now being constructed at this place [Shreveport] by Messrs. Moore & Smoker.
>
> Int. 1st. Are the boilers of said vessel sufficiently braced?
> 2nd. Has she the necessary "mud valves"?
> 3rd. Should the small engine for working the capstan be overhauled?
> 4th. Are there any pumps for pumping out the ship?

5th. Do the "exhaust valves" of the engines require grinding?
6th. Are the steam pipes so short as to require what is termed a "slip" or "expansion" joint?
7th. Is the Main Shaft sufficiently strong, and is it as good as could be done here?
8th. Are the joints of the Steam pipes sufficiently tight to prevent leakage?
9th. Do the cocks or pipes for supplying the boilers with water leak badly or is there any danger to be apprehended that said pipes will break or burst?
10th. Is the cock or pipe for supplying the deck pump with water necessary, and should the water for supplying said pump be furnished by the cocks and pipes that supply the boilers?[73]

These questions reveal a fairly complete picture of a Confederate ironclad's typical steam machinery components, but because the *Missouri* was built using riverboat parts, there were necessarily differences between it and the harbor ironclads of the coasts. In any case, Carter apparently received satisfactory answers, for his ship was in full operation by autumn 1863.

Letters pertaining to the *Missouri*'s steam plant in actual service began on September 19 when Carter wrote to Mallory, "The rate of speed of the 'Missouri' is 5½ miles per hour [without guns]." Carter was optimistic that even with the weight of guns, which he had great difficulty in procuring, the *Missouri* would steam better once it was on an even keel. He continued to find and fix small problems in the meantime: "The main bilge pumps are not as complete as should be for want of pipe" and "a small boiler and engine has been placed in the fire room which works a steam fan, steam capstan and steam pumps."[74] The *Missouri* was one of only six Confederate ironclads (with the *Atlanta*, the *Tennessee*, the *Tuscaloosa*, and the *Huntsville* completed, and the *Mississippi* uncompleted) known to have had forced draft fans installed in an effort to improve steaming qualities. It appears that the *Missouri*'s speed was in fact increased by this addition.

Carter's greatest difficulty in operating on the Red River was obtaining fuel. Many of his letters to his superiors beginning in December 1863 relate to gathering supplies of wood. On December 2 he wrote, "I would be glad . . . [to] have a lot of pine knots hauled to the river at some point below Alexandria for the use of the 'Missouri'" and explained further in another letter the same day, "Coal is not to be had and wood must be used which will be supplied by the 'Cotton' [a wooden paddlewheeler and the *Missouri*'s tender]."[75]

The *Missouri* remained trapped for nearly a year on the upper Red River by obstructions, low water, and lack of fuel. It could steam back and forth only occasionally. On September 20, 1864, Carter wrote once again about his lack of adequate fuel and the important services of the steamer *Cotton*: "The 'Missouri' can only carry wood for one day and should it become necessary to use her [in battle] the services of the Cotton' will be indispensable."[76] Unfortunately for its eager commander, the *Missouri* never saw action and waited out the waning days of the Civil War near Alexandria, Louisiana.

In late March 1865 Carter's last letters pertaining to the functioning of his ship's machinery reveal that despite all hardships, the *Missouri* performed well. Whether this would have been the case under the trying circumstances of battle is open to question, but it is apparent how much Carter attended to the needs of his ship. He especially had to maintain the steam pumps, for the *Missouri* leaked excessively because of green wood used in its construction. On March 24, 1865, with the war all but ended in the east, Carter wrote to one of the engineers who participated in installing the *Missouri*'s machinery: "Will you do me the favor to cast and fit a sliding valve for the steam pump of the 'Doctor Engine' on board the Steamer 'Missouri'?"[77]

Despite problems with leaks, the *Missouri* continued to function well until the end. The ship was finally forced to surrender in June 1865. With two months of idleness yet to go, on April 5, 1865, Carter summed up the functioning of his ironclad's surprisingly reliable power plant: "The performance of the 'Missouri' exceeded my expectation. Her speed is much greater than I expected. She has made as much as ten miles per hour with a current of from two to three miles. Her machinery works well."[78] This last statement is perhaps one of the highest compliments to steam engine operations in any Confederate ironclad.

The ironclad *Missouri* was the last Confederate vessel to surrender in home waters, on June 3, 1865.[79] Almost two months before, Lee had surrendered at Appomattox, and only a few remnants of Confederate forces were left anywhere. Union control of the Mississippi River and the relative isolation of the western frontier meant that the Shreveport and Alexandria areas on the Red River were surrendered last. Carter's *Missouri*, surrendered intact and operational, was a valuable prize. Accordingly, a survey was made of the ironclad.

The survey revealed the *Missouri*'s machinery specifications. The single-cylinder inclined poppet valve engines each measured 24 inches in diameter and 7.5 feet in stroke and were fitted to turn a single paddlewheel 22.5 feet in diameter and 17 feet wide. A major design fault was that the top 8 feet 4 inches of the wheel were exposed above the casemate top. Powering the ship were four horizontal double-flue boilers 26 feet long and 40 inches in diameter with flues 15 inches in diameter. The report also noted one doctor engine, one donkey boiler 8 feet long and 26 inches in diameter, and two donkey engines: one for running the capstan and the other for turning a fan blower 3 feet in diameter.[80]

A look at the preserved City class ironclad USS *Cairo*'s propulsion system reveals the same layout: the boilers were placed in the hold for protection, and the inclined engines extend up onto the gundeck to turn the wheel. Although the *Cairo* had five boilers, they were the same high-pressure horizontal type that most riverboats had and even had similar dimensions to the *Missouri*'s. They each measured 36 inches in diameter and 26 feet long with a 7.5-inch diameter flue. The two cast-iron engines were the typical riverboat type and also closely resembled the *Missouri*'s: they were mounted at a 15-degree angle and were 22 inches in cylinder bore with a 72-inch stroke.[81] The *Cairo*'s preserved machinery is listed as a National Historic Mechanical Engineering Landmark.

Unlike the Union vessel, the *Missouri* was never sunk but instead was taken to Mound City, Illinois, and sold for scrap.[82] It is therefore quite fortunate that the *Cairo* survives, for it is the only preserved paddlewheel-driven Civil War vessel in existence.

Conclusion

The *Albemarle*, the *Neuse*, the *Fredericksburg*, and the *Missouri* may almost be said to be a second generation of diamond hull ironclads. They were substantially different from the *Tuscaloosa* and the *Huntsville* and much more successful. Although the ships themselves varied in design, the common elements built into these ships under Constructor Porter's plans made for a type of ironclad much better suited to the Confederacy's lack of skilled shipbuilding infrastructure and shallow waterways.

The machinery installations on the four diamond hulls discussed in this chapter differed widely in type and source, but all appear to have given good service. The engines and boilers on the *Albemarle* and the *Neuse* particularly represent a makeshift solution to the problem of powering ironclads built in isolated areas. Both ships were slow (particularly the *Neuse*), but few problems with the actual operation of the machinery were noted by their commanders. On the shallow rivers of eastern North Carolina, draft was the primary concern.

The *Fredericksburg*, on the James River in Virginia, was a representative of Confederate shipbuilding at the heart of its greatest industrial center: Richmond. Although much is known about the design of this ironclad, surprisingly little information has been found on its machinery—what its layout was and how it performed. In contrast, the *Missouri*'s machinery is well documented. Scavenged steamboat machinery served it well during its uneventful career, and unlike with many other ironclads, there is plenty of contemporary information on its operation and configuration. Indeed, the steamboat engines and boilers used on the *Missouri* emphasized yet again how effectively the adapted riverboat machinery could work for ironclads.

The diamond hulls were among the last Confederate ironclads completed. The later examples, particularly the *Fredericksburg*, incorporated design improvements gained from previous wartime experience but were not substantially altered from their basic design. In contrast, by the end of the Civil War there were several ironclads nearly finished which included not only unusual and potentially revolutionary design characteristics but also superior machinery installations. Four of these, for which substantial details are known, will be discussed in the next chapter.

NINE

UNCOMPLETED VESSELS

We did all that we could . . . and had made all the necessary arrangements for the completion of the vessel. . . . It is impossible to describe our feelings.
—Nelson and Asa Tift

The Confederacy was ultimately successful in completing about two dozen ironclads out of approximately fifty placed under construction.[1] Considering an almost universal lack of infrastructure for industrial shipbuilding, the shortages of manpower and materials, and Union control of many Southern ports by the second year of the Civil War, two dozen is an important achievement. The Confederates were able to steadily improve upon existing designs and turn out better-armored warships as the conflict progressed. It should be noted, however, that many of the most ambitious ironclad designs were for vessels never completed.

The details for many uncompleted ironclads are lacking, but four came especially close to being finished. Taken together, they also represent a sort of cross-section of cutting-edge ideas in Confederate ironclad construction. CSS *Mississippi,* CSS *Jackson,* CSS *Milledgeville,* and CSS *Wilmington* differed widely not just in intended purpose but also in size, shape, and home port location. The *Mississippi* and the *Jackson* were actually begun early in the war, whereas the *Milledgeville* and the *Wilmington* represent some of the last ironclad designs conceived by Constructor Porter; indeed, they were some of the very last ironclads put under construction.

It has been seen that as the war progressed and the Confederate naval industry geared up for the immense task of building and outfitting ironclads, the quality and power of the supplied steam machinery also increased. Many of the late-war vessels were fitted with engines of good quality, especially from the Columbus Naval Iron Works in Georgia. By 1865 that facility, under the direction of and Chief Engineer James Warner, was the most important site for the construction and supply of marine engines. Most of the unfinished ironclads at that time would have received their machinery from those works.

The machinery intended for, or in some cases actually installed on, the unfinished ironclads discussed here itself often represented the best of what Confederate engineers could (or at least hoped) to achieve. Since none of these systems were ever put in service, it can only be speculated how they would have functioned. This is also true of the nature of the ships themselves (especially CSS *Mississippi*, which was an unusual design in so many ways). At present there is reason to believe that all four vessels discussed here would have been formidable warships if completed; any weaknesses they may have had remain conjectural.

CSS *Mississippi*

No other Confederate ironclad has spurred so much continuing debate as the giant CSS *Mississippi* at New Orleans. That ironclad was a truly singular vessel, and both its unusual construction and powerful machinery installation drew much attention. The *Mississippi* was begun early in the Civil War and belongs with the early nonstandard design category of ironclads (such as CSS *Louisiana*) because it was intended for both offensive and defensive operations on the rivers and the sea. This warship was considered the Confederacy's ultimate weapon in early 1862, if it could ever be finished.

Construction of the largest and most powerful ironclad built in the South was begun at the same time as on the *Louisiana* at Jefferson City, Louisiana, just upriver from New Orleans.[2] Nelson Tift (1810–1891) and his brother Asa (1812–1889) were originally from Connecticut but settled and made a name for themselves in Georgia and Key West. Nelson was successful as a planter, politician, and business entrepreneur by the time of the Civil War; Asa was involved in ship repair, warehousing, and merchandising in Key West. Neither had any shipbuilding experience, but both were prewar friends of Secretary Stephen Mallory, so they submitted plans for an enormous vessel to be driven by three screws and built without curved surfaces. This conception was a result of the brothers' perceptive realization that the Confederacy lacked experienced shipwrights.[3] The *Mississippi* was, therefore, constructed more like a ferry or a flat than a ship.[4] A rough sketch drawn by the Tifts illustrates the design's unusual conception and basic premise (fig. 9.1).

Many prominent naval officers and experts were on hand to observe and comment on the *Mississippi*'s form, and the plans were apparently approved by Constructor Porter. Some of the country's most senior officers praised the Tifts' concept. The *Mississippi*'s intended captain, Arthur Sinclair (1810–1865) touted the "formidable ship, the finest of the sort I ever saw in my life," as a potential "terror of the seas" capable of "not only . . . clear[ing] the river of the enemy's vessels, but rais[ing] the block-

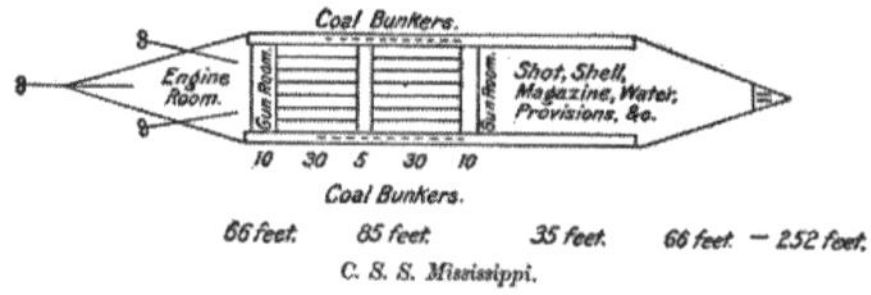

Figure 9.1. The *Mississippi's* general interior hull layout. The two rows of boilers are shown starting at the 85-foot mark, with the coal bunkers outboard of them against the hull sides. This allowed for a more efficient use of space but also provided another layer of protection to the unarmored hull in a ramming or wrecking event. Whether the brothers Nelson and Asa Tift realized that such a benefit existed is unknown, but along with the triple-screw arrangement it anticipated warship design decades in the future.

ade of every port in the South." Even Union Flag Officer David Farragut considered himself fortunate in not encountering a completed *Mississippi* during his campaign to seize New Orleans.[5]

There were a few others who doubted, though. Lieutenant Robert Minor, the commander of the Naval Ordnance Works in Richmond, observed, "By model she is waterborne in centre and near before and abaft the broadest beam but bow & stern are not waterborne sufficiently. . . . On the whole the work seems to be very reliable, but all depends upon the principle of construction which remains to be tested."[6] Minor, despite his experience and technical knowledge, seems to have been in the definite minority of opinion regarding the *Mississippi*. It is uncertain to what extent the engineers and mechanics actually working on the vessel agreed or disagreed with his opinions. The citizens of New Orleans certainly believed the ironclad to be an invincible weapon that could deliver their city from the Yankee aggressors gathering in the Gulf of Mexico. The Tifts found no lack of support for their ambitious project, both in the Crescent City (New Orleans) and in Richmond.

The first plank of the *Mississippi* was laid on October 14, 1861, but construction lagged despite the simplicity of the ship's design and its location adjacent to the Confederacy's largest shipbuilding center.[7] Among the most significant reasons for the delay was the *Mississippi's* sheer size: 240 feet long between perpendiculars, 58 feet in extreme beam, and 15 feet in depth of hold.[8] The length was increased by 20 feet during the course of construction when it was found during a consultation with experienced engineers that the existing boiler layout had insufficient grate and fire surface area to effectively propel the ship. The *Mississippi's* original design called for "11 boilers 32 feet long and 42 inches in diameter, 2 return flews [*sic*], with mat [mud] drum 24 inches in diameter, steam driver [steam drum] 30 inches in diameter, about 40 feet long." This already large system had to be upgraded to 16 double-flue boilers 42 inches in diameter and 30 feet long, capable of generating approximately 1,500 HP.[9]

The *Mississippi*'s triple 11-feet-in-diameter propeller arrangement was one of the first such systems ever built and required powerful engines of 36-inch bore and 24-inch stroke turning at a maximum of 125 RPM.[10] The stroke length was later increased to 30 inches during the revision of the boiler layout.[11] The engines were high-pressure horizontal direct-acting and were optimistically projected to drive the *Mississippi* at 14 knots.[12] Both boilers and engines, in addition to two doctor engines, two blowers, and two steam pumps, were constructed by Patterson Iron Works (formally known as Jackson & Company) of New Orleans.[13]

The size of the *Mississippi*'s machinery and propulsion components caused great difficulty for the manufacturers. The giant center propeller shaft had to be forged by Tredegar Iron Works in Richmond. No foundry in New Orleans could produce the 50-foot-long wrought-iron piece required.[14] In response to this dilemma the Tifts and Secretary Mallory contracted with Tredegar to forge the shaft from the burned-out steamer *Glen Cove* (sometimes called the *Glencoe*). The piece was painstakingly removed from the wreck, but no steam hammer in the Confederacy was capable of working such a large shaft, which had been built in the Northern states. Tredegar therefore had to resort to reworking it by hand.[15] Two roughly 25-foot-long sections were eventually completed with some difficulty. Fifty men labored night and day for two months to complete the work, and a special railroad car was built just to transport the immense piece of iron to New Orleans. Much time was lost in the process, and the shaft and center screw had just been installed when the *Mississippi* was destroyed.[16]

Forging the two wing shafts was also an odyssey. They were ultimately made by Clark & Company of New Orleans. Finding no complete shafts in the area and "no parties [there] . . . competent to make it" (i.e., no one apparently had the proper equipment), the Tifts contracted with Ward & Company of Nashville, Tennessee, to make the shafts. Negotiations with Ward about proper furnace and hammer equipment along with haggling over prices continued for some time but came to nothing. The Tifts next turned to Clark & Company which agreed to do the work. Clark then constructed a large steam hammer and furnaces in a new building for the purpose.[17] Further negotiations with Leeds & Company of New Orleans resulted in an agreement to finish the shafts once forged, and progress was finally made. The shafts were 9 inches in diameter and about 40 feet long.[18]

Manufacture, delivery, and installation of these pieces proved to be one of the most difficult aspects of the *Mississippi*'s construction. Nor was progress on the engines themselves turning out to be quick and simple. Under the advisement of a noted agent of Tredegar Iron Works, E. M.

Ivens, and according to reviews of the machinery layout by Chief Engineer James Warner, many changes were necessary to strengthen the engine component castings. This included most major parts, such as the piston rods, cylinders, heads, bedplates, main pillow block bolts, crank pins, and eccentrics.[19]

Patterson Iron Works agreed to furnish the engines and boilers after a consultation with Leeds had failed because that company was already occupied in manufacturing the shafting for the *Arkansas* and the never-finished *Tennessee* in Memphis.[20] Leeds's asking price for constructing the engines of $65,000 plus a build time of no less than four months was too steep for a short construction time. The Patterson works in turn offered to construct the machinery for $45,000 plus a bonus if the work was finished in ninety days, and received the contract.[21] When the necessary augmentations and improvements to the boilers and engine castings were specified, Patterson added $20,000 in price for building the engines and $8,000 for the boilers.[22]

The delays occasioned by the machinery revisions, and especially making the shafting, delayed the *Mississippi*'s launch date until April 19, 1862, just days before the Union captured New Orleans. By the fateful night of April 24–25 the boilers, smokestack, engines, center shaft and propeller, and wing shafts were on board, nearly complete but not connected.[23] As a result, the Confederacy's greatest warship was burned to prevent capture, after several attempts to tow the hulk upriver had failed.[24] The Tifts had just days earlier estimated a completion date of May 1.[25]

The *Mississippi*'s loss was a huge blow to Confederate morale and especially to the people of New Orleans. So great was the faith placed in its abilities that the anger and shock caused by the ship's destruction nearly cost the Tifts their lives at the hands of a furious lynch mob. Most were convinced that only treachery by the Northern-born brothers could have caused the loss of the already legendary *Mississippi*.[26]

Overall, of any Confederate ironclad, the *Mississippi* was affected in its construction most catastrophically by shortages of workers and materials and the limited abilities of local industry (despite it being abundant and accomplished). Even though the giant warship was of simplified design, there was not enough time to finish it, since New Orleans was captured at the end of the first year of war. As Secretary Mallory realized, there was great manufacturing potential in the Crescent City, but it took too long to build up to full wartime production. In the end the New Orleans firms' inability to cope with increased industrial demand at the beginning of the Civil War resulted in the premature loss of the city and the Confederacy's two most powerful ironclads, the *Louisiana* and the *Mississippi*.[27]

Figure 9.2. Original photograph of the *Jackson* taken soon after its launching, December 1864. The ship floats high in the water without armor, machinery, or ordnance; the opening left in the front of the casemate is presumably for the installation of the latter. Note the ram bow, the gunport configuration (three per side and two each on the forward and aft casemate faces), and the small pilothouse. An awning spread on the aft flush deck at the left nearly obscures the wooden gunboat CSS *Chattahoochee* that is moored behind the ironclad.

CSS *Jackson*

Despite the immense contribution of Columbus, Georgia, to ironclad construction and machinery assembly, only one ironclad was actually built there. It would seem that such a project at such a place would have all the advantages that other ironclad construction sites did not, but from the start nothing seemed to go right. Construction of the Columbus ironclad, intended to guard the Chattahoochee River as far as Apalachicola Bay in Florida, became a painstaking effort that lasted for most of the Civil War. In many ways the construction of this one ship became the primary purpose (almost a pet project) of Lieutenant Augustus McLaughlin and Chief Engineer James Warner in their administration of the Columbus Navy Yard and of Columbus Naval Iron Works, respectively.

CSS *Jackson*, also commonly called the *Muscogee*, was originally intended to be a sister design to the *Missouri*. It was therefore the second (and last) diamond hull centerwheel ironclad to be laid down. The ship was built next to Columbus Naval Iron Works but suffered from continuous construction problems and became the most altered ironclad built in the Confederacy. These changes were not represented in the historical record but are now well documented archaeologically because of the ironclad's recovery in the early 1960s.[28]

The *Jackson* was originally laid down in December 1862 but was not successfully launched until after its major reconstruction into a twin-propeller vessel was recommended by Constructor Porter. Porter's visit was the result of a failed launch attempt on January 1, 1864, in which it was revealed that the *Jackson* was too heavy to float off the blocks even with the bow partly afloat. This disturbing fact, coupled with a realization of the design's unsuitably deep draft, called for reconstruction and further delayed completion. The *Jackson* was not successfully launched until December 22, 1864. A photograph of the ship was taken soon afterward (fig. 9.2).[29]

The *Jackson*'s original design conformed closely to Porter's centerwheel ironclad plans but had to be extensively lengthened at the stern for twin-screw propulsion. This ultimately resulted in a vessel approximately 225 feet long overall, 59 feet in extreme beam, and 7 to 8 feet in loaded draft.[30] The paddlewheel machinery previously obtained for and installed in the ship had to be removed.[31] Although it is unknown what was done with these engines afterward, their source is known—the 263-ton sidewheeler *Time*, built in 1860 at Louisville, Kentucky.[32]

The *Time* was the last steamboat to enter service on the Apalachicola-Chattahoochee River system before the Civil War. It later came under Confederate control and may have undergone a name change to the *Bradford*.[33] Conflicting accounts of the ship's activities render a timeline difficult to establish. There is even evidence that the *Time* and the *Bradford* were separate boats, but as late as March 1862, at least one of them was still operating.[34]

At some point the *Time* was dismantled and its machinery appropriated for the ironclad building at Columbus. Everything was actually placed in the *Jackson* and mostly put together before the need for rebuilding became apparent.[35] The provenance of the *Time*'s power plant remains unknown, as does its fate after removal to make way for the Columbus-built propeller engines of the *Jackson*'s second incarnation.

A recently discovered drawing of the *Time*'s engines in a private collection, labeled "Muscogee's Engines," reveals the approximate cylinder diameter to be 30 inches, with a stroke of about 9 feet. The engines appear to be the usual riverboat type, and presumably so were the boilers. There were four of these on the *Time*, and they seem to have been kept in the *Jackson*'s second incarnation as a propeller-driven vessel.[36] No information on the boiler dimensions has been found.

The second set of engines supplied for the *Jackson* were purpose built by the adjacent Columbus Naval Iron Works and were of the typical horizontal single-cylinder direct-acting type, measuring 36 inches in bore and stroke.[37] Each engine independently turned a screw 7.5 feet in diameter and mounted on a shaft 6.75 inches in diameter. The mounting timbers of the engines are preserved today, as are the ends of the shafts, and both can be clearly seen when observing the remains from the gallery level of the *Jackson's* display hall at the Port Columbus Civil War Naval Museum.

Although the engines of the *Jackson* no longer exist, those of the wooden Columbus-built gunboat CSS *Chattahoochee* are preserved at the museum. These represent the only extant examples of the preferred engine type for Confederate ironclads and measure 28 inches in bore, 34.5 inches in exterior diameter, and 20 inches in stroke. They are similar in virtually all respects to the horizontal engines installed on the *Jackson* and several other ironclads.

The *Jackson*'s late launch date resulted in its never being finished, and the ship was set on fire by invading Union troops on April 17, 1865.[38] After Sherman's infamous March to the Sea cut its way across northern Georgia in late 1864, other federal forces under General James H. Wilson worked their way across Alabama from west to east, mopping up the last areas of Confederate resistance as they went. They finally reached Columbus on the night of April 16–17. In the midst of the resulting battle, much of the city was laid waste from artillery barrage and by the retreating Confederates. The naval works were among the last to go: cavalry under Union General Edward F. Winslow captured the facility along with the still incomplete *Jackson*, made a brief inventory, then systematically destroyed the Confederacy's last naval-industrial complex.[39]

The *Jackson* was torched and its moorings were cut. Drifting slowly with the current, the burning ironclad floated downstream, past the place where the gunboat CSS *Chattahoochee* lay afire from its own crew, for 30 river miles before grounding. The *Jackson* was afterward deemed not worth salvaging because of extensive damage caused by the fire, but in 1910 the US Army Corps of Engineers dredged around the wreck and removed the machinery.[40] The recovered remains have provided significant information on the construction of diamond hull ironclads as well as on the *Jackson*'s ill-fated design changes.

CSS *Milledgeville*

This one-of-a-kind vessel was a late-war design intended to be an improvement over the earlier standard hull types such as the *Richmond* class. As such, the *Milledgeville* had a similar hull structure to that of earlier ironclads but incorporated significant design changes. Fortunately, these changes are shown in extant plans of the ironclad bearing Constructor Porter's signature (fig. 9.3). The *Milledgeville* was a twin-screw standard hull ironclad laid down in February 1863 by Henry Willink, the constructor of CSS *Savannah* and a prominent and skilled local shipbuilder.[41] The new

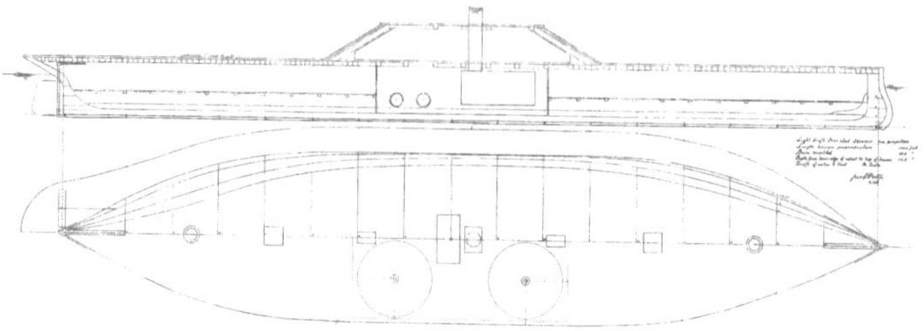

Figure 9.3. Porter's plans of the *Milledgeville*, showing details of engineering and gun deck arrangements. The final configuration of *Milledgeville*'s power plant was reportedly nearly identical to other Columbus-built examples. Note the lack of a ram bow and the circular features on the bow and stern flush decks shown in the plan view at the bottom. These are believed to be cowl ventilators.

ironclad was 175 feet in length between perpendiculars, 185 feet in overall length, 35 feet 2 inches in molded beam, and 48.5 feet in extreme beam. It featured the late-war improvement of the flush deck design of the *Columbia*, which continued to flare up to main deck level instead of turning back at the waterline. This allowed for a wider main deck and more room for multiple pivot guns in the shortened casemate, which was covered with 6 inches of armor. Most important, the ship had a reduced draft of 9 feet.[42]

The *Milledgeville* plans also show the general machinery arrangement of the ship, although its final configuration may have differed slightly. Unfortunately, like those of most other Confederate ironclads, the plans show little actual detail of the engines and boilers. Two circles scaled to approximately 30 inches in diameter representing single-cylinder horizontal engines are shown in the starboard plan view along with a boiler 14 feet long and 6 feet in diameter. Dimensions taken off the plans show that the *Milledgeville* had two boilers. Twin propellers 7 feet in diameter are shown at the stern. All the machinery was built by Columbus Naval Iron Works and should have given good service.[43]

The *Milledgeville*'s completion was severely delayed, like many other Confederate ironclads. Willink's prior commitments and design changes to the ship while under construction were the biggest factors.[44] The *Milledgeville* was launched in the early fall of 1864 and was nearly complete when it was burned to prevent capture as Union forces under General Sherman reached Savannah on December 21, 1864.[45] The ship was anchored and ready for towing out of harm's way, but there were no vessels available to help. After removing some stores and readying the hulk for maximum destructive effect, the Confederates set fire to the ironclad; it was soon sunk.[46]

The *Milledgeville*'s engines and most of its armor were installed by that time but were salvaged by the US Army Corps of Engineers in the

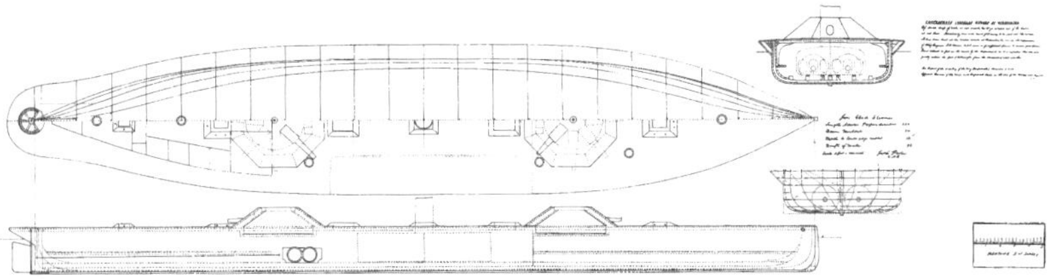

Figure 9.4. John L. Porter's plans of the *Wilmington*, showing hull construction and some engineering detail. The four boilers are amidships, with the forward and aft pairs back-to-back to form a single smokestack. As in the *Milledgeville* plans shown in the previous figure, there is a lack of a ram bow along with small circles shown in the plan view of the ship's main deck that represent cowl ventilators. The small sloped hatches with viewing ports just forward and aft of the small casemates are the pilot stations.

1890s.[47] The current condition of the wreck is unknown—future archaeological surveys of the site may uncover new details.

CSS *Wilmington*

Almost to the end of the Civil War the vital port of Wilmington, North Carolina, remained open to blockade-runners. The Confederate naval squadron there, however, was one of the poorest and most beset by bad luck during the conflict. It was vitally important that a suitable new ironclad be built to defend the Cape Fear River approaches.

The design of CSS *Wilmington*, informally named for the city (an official naming ceremony never seems to have occurred), reflected both lessons learned in ironclad design and the unique nature of its probable mission. The *Wilmington* was one of the very last Confederate ironclads to approach completion. It was never launched and was burned on the stocks as Union troops finally reached the Confederacy's last open port in February 1865.[48] Fortunately, enough information—the original plans (fig. 9.4) and an engineering diagram (fig. 9.5), both located at the Port Columbus Civil War Naval Museum—exists on this unique vessel for an accurate reconstruction to be made.

The *Wilmington* was designed by Constructor Porter as an effective replacement for the wrecked *Raleigh* and rotten *North Carolina* in May 1864. It was laid down at the Beery shipyard on Eagles Island where the *North Carolina* had been built, and some of the armor from that ship was actually used in the *Wilmington*'s construction. The ship's design reflects a possible intention for use as a blockade breaker: it had very sleek lines, with a length between perpendiculars of 224 feet and a molded beam of 34 feet. An overall length of 233 feet 4 inches, an extreme beam of 41 feet, a depth of hold of 12 feet, and a draft of 9.5 feet resulted in an ironclad not only very different from the *Richmond* class vessels but also superiorly

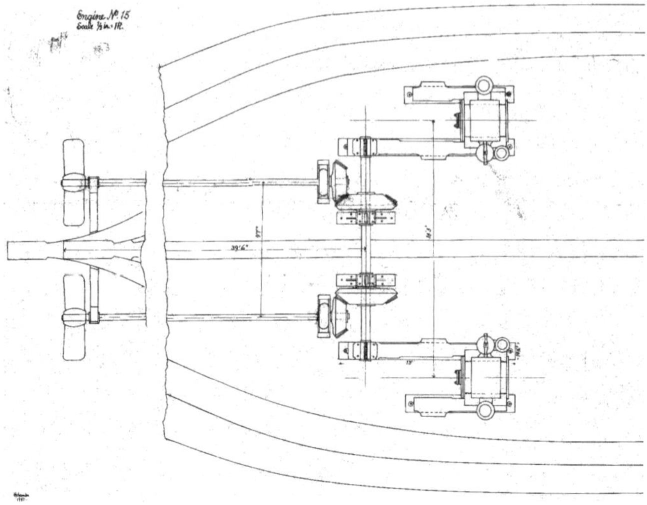

Figure 9.5. Bob Holcombe's 1981 redrawing of an engineering diagram of a twin-screw vessel fitted with Engine No. 15. The dimensions and layout indicate that this vessel is the *Wilmington*. The two horizontal engines with overhead valve chests are on either side of the keel assembly with their cylinders facing forward. Note that the piston and connecting rods are not shown. These would have turned the transverse mounted idler shaft with its two large gears, which turned the two smaller gears meshed with them at 90 degrees. These smaller gears were directly attached to the propeller shafts. Though seemingly complicated, this machinery installation potentially offers higher revolutions per minute for the twin screws.

adapted to its operating environment. A high design speed and low, rakish appearance with twin single-gun casemates only adds to the hypothesis that the *Wilmington* was to be used for something other than harbor defense purposes.[49] A severe lack of iron for armor at that late stage of the war certainly cannot alone explain these sleek design features.

The engineering diagram that shows this ironclad's high speed machinery is also quite informative. The *Wilmington*'s power plant was built by Columbus Naval Iron Works but never delivered.[50] The ironclad was to be equipped with four triple furnace boilers, each approximately 16 feet long, 6.5 feet high, and 8 feet wide, powering two high-pressure single-cylinder horizontal engines 28 inches in bore, 24 inches in stroke, and placed fore and aft in the hull. Each engine was connected to the propeller shafts by two gears, one 4.5 feet and the other 3 feet in diameter, which allowed for potentially greater speed through higher revolutions.[51] Twin propellers each 8 feet in diameter were prevented from independent maneuvering by the use of an idler shaft. Since the advantages of independently operating propellers were generally known by the 1860s, the reasoning behind this design handicap is a mystery.

The engines and boilers of the *Wilmington* never left Columbus and were captured in April 1865 along with many other materials.[52] In many ways it is a shame that this ironclad and other late-war designs were never completed, because not only did they reflect significant design improvements, their purpose-built machinery also should have been among the best made by either side at that late stage. The *Wilmington* perhaps emphasizes best of all the incredible growth and development of the Confederate naval industry during the Civil War and at the same time that system's failure.

Conclusion

The uncompleted ironclads discussed in this chapter represent most of all the continual growth and development of the Confederate naval industry during the Civil War. In the end the system was forced to adapt to existing contingencies rather than the ideals of designers and administrators. The *Mississippi*, the *Jackson*, the *Milledgeville*, and the *Wilmington* show this process in their different and increasingly improved designs.

The *Mississippi* was the first laid down and represented the early war's conflict of purpose and the idealism of ironclads as an ultimate weapon. Its huge size and complex machinery installation were ultimately the cause of its premature destruction in a nearly finished state. In contrast, the *Jackson*, the next one laid down, reflected Constructor Porter's definite purpose in shallow-draft diamond hulls for riparian operations, but it underwent so many design modifications that it became virtually a different ship from the one originally conceived. Its construction at the great industrial center of Columbus did not aid in its completion, although unlike most ironclads, suitable engines were readily available for it.

The *Milledgeville* and the *Wilmington*, unlike the *Mississippi* and the *Jackson*, were both late-war designs that benefited from previous experience in construction, machinery operation, and combat. They can both be considered pinnacles of Confederate ironclad design. Both seem to have been well constructed, suitable for use both in harbors and outside against the blockade, and, above all, provided with good machinery from Columbus. Although these two ironclads never amounted to anything, they represent most prominently how much Southern naval construction efforts improved in four years of war. The Confederate marine steam industry was at the forefront of this improvement.

TEN

RESULTS AND OVERVIEW

So far as human ingenuity contending against meagre facilities could accomplish.
—Commander John N. Maffitt, Confederate States Navy

The general failure of the Confederate navy to effectively construct quality marine steam machinery until the Civil War's end played an important role in the outcome of its shipbuilding programs. The ironclad program was the most important of these in resources allocated and sheer numbers. By 1865 approximately fifty ironclads had been laid down or contracted for, and of these twenty-three were placed into service.[1] These numbers far exceeded those of the wooden gunboats and other small craft actually constructed (compared to the vast numbers initially intended) in the early years of the war. As a result, the ironclads demanded a huge amount of material and technical know-how that was often not available in time or of usable quantity. It is all the more remarkable, then, to consider the fact that any suitable steam machinery was obtained at all.

A significant lack of experience in machinery manufacturing was just one of many hurdles that Secretary Mallory and the fledgling Confederate navy had to overcome in 1861. As former Confederate officer Charles H. McBlair stated:

> In the entire Confederacy there was not a reliable establishment . . . where a first-class marine engine could be constructed. Three rolling-mills only existed in the whole country, two of which, in Tennessee, were unavailable owing to the disturbed condition of the districts in which they were located, and the third, at Atlanta, Georgia, had to undergo extensive alteration, involving serious delay, before it could undertake the heavy work of the navy. The private machine-shops and foundries resorted to by the government in the absence of public establishments of the kind were everywhere more or less crippled by the loss of operatives, who had enlisted in the land service.[2]

In essence, the Confederate navy started with nothing but a set of experienced officers resigned from US service, but it pioneered the organizational and technical basis of a modern navy by 1865. This included not only an ironclad fleet but also mine, submarine, and commerce raiding warfare. All these elements reflected the revolutionary new nature of war at sea.[3]

Several industries greatly scaled up production and achieved significant successes in heavy manufacturing in order to provide these modern weapons to the Confederate cause. Tredegar Iron Works, Shockoe Foundry (later the Confederate Naval Works), the Charlotte Navy Yard, and Columbus Naval Iron Works were the greatest of these. The latter especially contributed to the manufacture of steam machinery in the Confederacy, and by the war's end several complete and well-built sets had been delivered for installation in ironclads.

The various sets of machinery supplied to the Confederate ironclads ultimately reflected a combination of sources. Single-cylinder vertical-inverted direct-acting tugboat engines, high-pressure inclined riverboat systems, and horizontal single-cylinder direct-acting engines were the three most important groups utilized. The purpose-built specimens represented by the horizontal direct-acting type were overwhelmingly the best of these. This was largely because each set of new engines was specifically intended for a certain ironclad from the outset, and it was also a result of refinements in manufacturing during the war years. The secondhand power plants were usually unsatisfactory, although it appears that many of the scavenged riverboat systems, such as those on the *Missouri* and the *Tennessee*, performed surprisingly well.

Many ironclads, especially those of the *Richmond* class, utilized former tugboat engines. These were often the most temperamental sets of machinery because of hard usage, age, and overstressed working when placed in a heavy ironclad. Those vessels so equipped were invariably underpowered. This was not necessarily a crippling problem, since virtually all the Confederate ironclads built after 1862 were intended for harbor defense.[4] Lack of speed did render ramming more difficult in many cases, perhaps best emphasized by the January 1863 sortie of the *Chicora* and the *Palmetto State* from Charleston Harbor. In at least one case, that of the *North Carolina*, a weak engine indirectly resulted in sinking by shipworm damage. Often the purpose-built boilers and auxiliary equipment manufactured by local industry were the best part of the power plants of the ironclads equipped with tugboat engines.

Civil War–era steam boilers came in a variety of types, like the engines they powered, but on the Confederate ironclads the most common were the low-pressure horizontal firetube type and the western rivers–style horizontal high-pressure flue type. Both were ideally suited to their respective purposes, although the high-pressure type was generally more prone to explosion. Boilers were fairly easy to build, unlike steam engines, and the Confederates had no problem procuring them for ironclads. Several local firms in Wilmington, Charleston, Savannah, and elsewhere had the capability of producing even the newer designs, rendering effective construction times and local deliveries. As a whole, the boilers and auxiliary equipment of the Confederate ironclads were standard for the day and much less troublesome to manufacture and operate than the engines.

Propellers and paddlewheel propulsion existed simultaneously by the Civil War, sometimes even on the same ship (such as the *Louisiana*). Propellers were only just beginning to supersede paddlewheels, and only then mostly on oceangoing vessels. Despite its seemingly vulnerable and antiquated form, the paddlewheel continued to be a viable and effective means of propelling shallow-draft coastal and riparian vessels well into the twentieth century. Regarding the Confederate ironclads, no truly effective comparison can be made between those vessels fitted with paddlewheels and those with propellers, because of differences in construction and operating environment and the general prevalence of screw propeller designs. In fact, the only ironclads driven by paddlewheels that were finished (the *Baltic*, the *Louisiana*, the *Missouri*, and the *Nashville*) reflected different operating philosophies and layouts. Of the paddlewheel ironclads, the *Baltic* and the *Nashville* were sidewheelers and the *Louisiana* and the *Missouri* were centerwheelers. Only the *Missouri* can be considered somewhat of a success. This was largely a result of design rather than engine type, since the western riverboat power plants provided for these vessels generally functioned well even under the strain of guns and armor.

The vast majority of the Confederate ironclads were screw-propelled. This was mostly a result of the desire to protect the ship's drive system from battle damage and ramming. As was emphasized spectacularly by the *Nashville*, paddlewheels, especially sidewheels, were not suited for coupling to the angled casemate and were greatly exposed. The submerged, recessed nature of the propeller allowed for greater protection, although its more complicated emplacement and machinery connections often caused trouble in those early days of steam propulsion. In addition, the use of twin propellers on shallow-draft designs like the *Albemarle* potentially allowed for greater maneuvering ability. Often this layout was handicapped for unknown reasons by complicated gearing or coupling to prevent independent turning of the screws. Propellers were also more difficult to manufacture and maintain than paddlewheels, but no significant problems seem to have been encountered, even with ironclads such as the *Neuse* operating in extremely shallow water.

It has been commonly emphasized since the Civil War that all Confederate ironclads were inherently defective. This is not true—although their machinery often left much to be desired, the Confederate ironclads were generally strongly built and adequately armed. Even the use of green wood in many cases did not negate their strength in battle; for example, the *Louisiana* withstood the heavy guns of Farragut's fleet with only superficial damage, and the *Arkansas* was able to survive the gauntlet of the combined Union fleets above Vicksburg. It must be remembered that Union officers made every effort to capture a Confederate ironclad when possible. Two, the *Tennessee* and the *Atlanta*, served effectively as Union vessels later in the war.

This work has resulted in a compilation of data on twenty-seven Confederate ironclads and their steam machinery—twenty-three finished and four nearly finished. Here, presented for the first time, are detailed descriptions of machinery types commonly used on ironclads of the Civil War and a view of those vessels in action. The firsthand engineering accounts from Confederate officers have proved to be quite rare and represent only a handful of ironclads, but they offer an invaluable insight into the lives and dedication of these early steam pioneers. Without their services, which in the agrarian Confederacy were scarce to begin with, Mallory's new state-of-the-art steam navy could not have functioned. The testimony of these men also offers valuable insight into the defects of their charges and aids in a final summation.

Trends in building and purchasing marine steam machinery, and also the testimony of engineers, are clearly revealed by this work. Only one large gap remains: a conclusive appraisal of Confederate Engineer in Chief William P. Williamson's career and contributions. Unfortunately, this may never be possible because of the lack of information relating to his service. The overall conclusions have not changed regarding Confederate marine steam machinery, particularly on ironclads, and this study has only corroborated them. The final summation is best made by distinguished naval historian Dr. William N. Still Jr.:

> Of all aspects of the Confederate shipbuilding program, the marine engineering industry was by far the most backward, as it was in the United States as a whole. There was little scientific study of engineering before the war; construction of marine engines was frequently by rule of thumb. This was true throughout the war, in the North as well as in the South. All parts were generally made by hand,

> and although the concept of interchangeable parts for engines was quite common in the arms industry, it had not reached the marine engineering industry. In a field where all knowledge was empirical, it was inevitable that the Confederacy's first engines and other machinery would be inefficient. Nevertheless, Confederate marine engines, like the vessels themselves, improved with time, and by the end of the war a few satisfactory engines were being produced.[5]

Credit was due to the dedication and drive of the men in charge of the Southern industries and their leadership in Navy Secretary Mallory. They made incredible gains in the face of overwhelming odds not only to industrialize during wartime but also to render a navy from scratch. Although the Confederate navy and its ironclads played mostly a defensive role, several major ports such as Wilmington, Charleston, and Mobile were kept open until nearly the war's end by the presence of steam-powered armored vessels. Thus the crucible of war in the United States paved the way for the birth of a modern navy: steam-powered sail-less warships, mines, submarines, and commerce raiding—all hallmarks of those great naval actions beginning half a century later in World War I. It is remarkable to consider the fact that an agrarian nation was able to make such gains in four short years and just as remarkable that the opposing side was able to improve upon and counter those gains, leading to an explosion of new technical developments at sea. Just twenty short years after the Civil War, those continuing developments allowed the United States Navy to compete successfully on the world stage for the first time with the launch and shakedown of its first modern seagoing armored cruisers.

Although it is not widely appreciated today, or even was in its own time, the Confederate navy was responsible in large part for many new technical developments. The ironclad program is probably the best known of these. Secretary Mallory's vision allowed for the creation of what may tentatively be labeled the first all-modern navy. The Confederate navy represented an impressive achievement, especially in the development of the naval-industrial complex and marine machinery manufacture, despite lasting only four years. These facts were certainly not lost on Mallory, who despite his deficiencies can be considered a visionary and the best man for the job.[6] My final statement on the Confederate navy is in his words, written to former Confederate Commandant of Midshipmen James H. Rochelle on May 21, 1867: "I am satisfied that, with the means at our control and in view of the overwhelming force of the enemy at the outset of the struggle, our little navy accomplished more than could have been looked or hoped for; and if I have ever felt any surprise connected with its operations, it was that we accomplished so much."[7]

APPENDIX

Confederate Ironclad Steam Machinery Specifications

Conversions

1. CSS *Manassas* (ex–*Enoch Train*)

- Engines: Two single-cylinder inclined direct-acting; 36 inches in diameter by 32 inches stroke; low-pressure condensing but capable of high-pressure working; manufacturer, Loring's City Point Works, Boston, Massachusetts.
- Boilers (original): Probably one or two horizontal return flue; dimensions unknown; low pressure; manufacturer, Loring's City Point Works, Boston, Massachusetts.
- Boilers (as converted): Probably three or four horizontal return flue; dimensions unknown; high pressure; manufacturer unknown.
- Propulsion system: Two propellers, 9 feet in diameter; independently turning; manufacturer, Loring's City Point Works, Boston.
- Vessel disposition: Wreck site buried in Mississippi River bottom near New Orleans, Louisiana, condition unknown.

2. CSS *Virginia* (ex–USS *Merrimack*)

- Engines: Two single-cylinder horizontal return piston rod (back acting); 72 inches in diameter by 36 inches stroke; low-pressure condensing; manufacturer, West Point Foundry, Cold Spring, New York.
- Boilers: Four Martin vertical watertubes, three furnaces each, 15 feet high by 14 feet wide by 12 feet long; low pressure; manufacturer, West Point Foundry, Cold Spring, New York.
- Propulsion system: One propeller, 17 feet 4 inches in diameter; Griffiths patent propeller with variable pitch; manufacturer, West Point Foundry, Cold Spring, New York.
- Vessel disposition: Wreck site buried under Craney Island dredge materials, Portsmouth, Virginia, condition unknown.

3. CSS *Baltic*

- Engines: Two single-cylinder inclined direct-acting; 22 inches in diameter by 7 feet stroke; high-pressure noncondensing; manufacturer, Lent, South & Shipman, New Albany, Indiana.
- Boilers: Four horizontal return flue; 24 or 28 feet long by 3 feet or 40

inches in diameter; high pressure; manufacturer, Lent, South & Shipman, New Albany, Indiana.

- Propulsion system: Two sidewheels, 29 feet in diameter by 8 feet wide; independently turning; manufacturer, Lent, South & Shipman, New Albany, Indiana.
- Vessel disposition: Sold by US Navy, Dec. 31, 1865; scrapped at later date.

Early Nonstandard Types

4. CSS *Arkansas*

- Engines: Two single-cylinder horizontal direct-acting; 30 inches in diameter by 30 inches stroke; high-pressure noncondensing; manufacturer, Union Foundry & Machine Shop or S. M. Coates foundry, Memphis, Tennessee.
- Boilers: Six horizontal, probably return flue; dimensions unknown; high pressure; manufacturer, Union Foundry & Machine Shop or S. M. Coates foundry, Memphis, Tennessee.
- Propulsion system: Two propellers, 8 feet in diameter; independently turning; manufacturer, Leeds & Company, New Orleans, Louisiana.
- Vessel disposition: Wreck site buried in Mississippi River bottom near Baton Rouge, Louisiana, condition unknown.

5. CSS *Louisiana*

- Primary engines: Two single-cylinder inclined direct-acting; 28 or 30.5 inches in diameter by 7 or 9 feet stroke; high-pressure noncondensing, taken from the river steamboat *Ingomar*; manufacturer, Roach & Long, Louisville, Kentucky.
- Secondary engines: Two probably vertical-inverted direct-acting; probably 18 inches in diameter by 18 inches stroke; probably low pressure, probably taken from the steam tugboats *Decatur* and *America*; manufacturer, probably Reanie, Neafie & Company, Philadelphia, Pennsylvania.
- Boilers: Five horizontal return flue; 33 feet long by 42 inches in diameter; high pressure, taken from the river steamboat *Ingomar*; manufacturer, Roach & Long, Louisville, Kentucky.
- Primary propulsion system: Two centerwheels, 27 feet in diameter by 19 feet wide or 20 feet in diameter (fore) and 24 feet in diameter (aft); manufacturer, E. C. Murray, Jefferson City, Louisiana.
- Secondary propulsion system: Two propellers, 5 feet in diameter; manufacturer, Kirk & Company, New Orleans, Louisiana.
- Vessel disposition: Wreck site buried in Mississippi River bottom near New Orleans, Louisiana, condition unknown.

6. CSS *Georgia*

- Engines: Two single-cylinder horizontal direct-acting; 24 inches in

diameter by 39.5 inches stroke; low-pressure condensing; manufacturer, probably Alvin N. Miller foundry, Savannah, Georgia.
- Boilers: One or two horizontal firetubes; approximately 18 feet long by 7.5 feet wide; low pressure, standard marine variant of locomotive boiler; manufacturer, probably Alvin N. Miller foundry, Savannah, Georgia.
- Propulsion system: Two propellers, 8 feet in diameter; manufacturer, probably Alvin N. Miller foundry, Savannah, Georgia.
- Vessel disposition: Wreck site in Savannah River near Savannah, Georgia, extensively surveyed, then excavated 2014–2015; conservation is ongoing.

Richmond Class Vessels

7. CSS *Richmond*

- Engine: One single-cylinder vertical-inverted direct-acting; dimensions unknown; low-pressure condensing, taken from steamer CSS *Arctic* (ex–USS *Arctic*, ex–Lightship No. 8); manufacturer, James T. Sutton Company of Philadelphia.
- Boilers: Two horizontal firetubes; approximately 10 feet long by 11 feet high by 6 feet 9 inches wide; low pressure; manufacturer, Tredegar Iron Works or Shockoe Foundry, Richmond, Virginia.
- Propulsion system: One propeller, 8 or 10 feet in diameter; manufacturer, Tredegar Iron Works or Shockoe Foundry, Richmond, Virginia.
- Vessel disposition: Wreck site buried in James River bottom near Richmond, Virginia, condition unknown.

8. CSS *Chicora*

- Engine: One single-cylinder vertical-inverted direct-acting; 30 inches diameter by 28 inches stroke; low-pressure condensing, taken from the steam tugboat *Aid* (aka the *Concord*); manufacturer, Reanie, Neafie & Company, Philadelphia, Pennsylvania.
- Boilers: Two horizontal firetubes; approximately 10 feet long by 11 feet high by 6 feet 9 inches wide; low pressure; manufacturer, James M. Eason & Brother, Charleston, South Carolina.
- Propulsion system: One propeller, 8 feet in diameter; manufacturer, J. M. Eason & Brother, Charleston, South Carolina.
- Vessel disposition: Wreck site buried in Charleston Harbor bottom, South Carolina, condition uncertain but at least some scattered components appear to remain.

9. CSS *Palmetto State*

- Engines: Two single-cylinder horizontal direct-acting; 30 inches in diameter by 26 inches stroke; low-pressure condensing, taken from CSS *Lady Davis* (the ex–steam tugboat *James Gray*, aka the *Tompkins*);

manufacturer, Reanie, Neafie & Company, Philadelphia, Pennsylvania.

- Boilers: Two horizontal firetubes; approximately 10 feet long by 11 feet high by 6 feet 9 inches wide; low pressure; manufacturer, Cameron & Company, Charleston, South Carolina.
- Propulsion system: One propeller, 8 or 10 feet in diameter; manufacturer, Cameron & Company, Charleston, South Carolina.
- Vessel disposition: Wreck site buried in Charleston Harbor bottom, South Carolina, condition unknown.

10. CSS *North Carolina*

- Engine: One single-cylinder vertical-inverted direct-acting; dimensions unknown; low-pressure condensing, taken from the steam tugboat *Uncle Ben*; manufacturer, Shepard Iron Works, Buffalo, New York.
- Boilers: Two horizontal firetubes; approximately 10 feet long by 11 feet high by 6 feet 9 inches wide; low pressure; manufacturer, Hart & Bailey Iron Works, Wilmington, North Carolina.
- Propulsion system: One propeller, 8 or 10 feet in diameter; manufacturer, Hart & Bailey Iron Works, Wilmington, North Carolina.
- Vessel disposition: Wreck site near Battery Island at Southport, North Carolina, partly surveyed, some small artifacts excavated and conserved.

11. CSS *Raleigh*

- Engines: Two single-cylinder horizontal direct-acting; 24 inches in diameter by 36 inches stroke; low-pressure condensing, probably originally for unfinished wooden gunboat at Washington, North Carolina; manufacturer, Tredegar Iron Works or Shockoe Foundry, Richmond, Virginia.
- Boilers: Two horizontal firetubes; approximately 10 feet long by 11 feet high by 6 feet 9 inches wide; low pressure; manufacturer, Clarendon Iron Works, Wilmington, North Carolina.
- Propulsion system: 1 propeller, 8 or 10 feet in diameter; manufacturer, Clarendon Iron Works, Wilmington, North Carolina.
- Vessel disposition: Wreck site in Cape Fear River near Fort Fisher, North Carolina, partly surveyed, some small artifacts excavated and conserved.

12. CSS *Savannah*

- Engines: Two single-cylinder horizontal direct-acting; 28-inch bore (34.5 inches in external diameter) by 20-inch stroke; low-pressure condensing, originally for unfinished wooden gunboat at Milton, Florida; manufacturer, Columbus Naval Iron Works, Columbus, Georgia.
- Boilers: Two horizontal firetubes, three furnaces each; 11 feet 2 inches long by 12 feet high by 7 feet 10 inches wide; low pressure; manufacturer, Columbus Naval Iron Works, Columbus, Georgia.

- Propulsion system: One propeller, 10.5 feet in diameter; manufacturer, Columbus Naval Iron Works, Columbus, Georgia.
- Vessel disposition: Wreck site in Savannah River near Savannah, Georgia, condition unknown.

Tennessee Class Vessels

13. CSS *Tennessee*

- Engines: Two adapted single-cylinder inclined direct-acting; 24 inches in diameter by 7 feet stroke; high-pressure noncondensing, taken from the river steamboat *Vicksburg*; manufacturer unknown.
- Boilers: Four horizontal return flues, one furnace under all; 24 feet long, other dimensions unknown; high pressure, taken from the river steamboat *Vicksburg*; manufacturer unknown.
- Propulsion system: One propeller, 12 feet in diameter; geared to engines; manufacturer, Charlotte Navy Yard, Charlotte, North Carolina.
- Vessel disposition: Sold by US Navy, November 27, 1867, scrapped at later date.

14. CSS *Columbia*

- Engines: Two single-cylinder horizontal direct-acting; 36 inches in diameter by 24 inches stroke; high-pressure noncondensing; manufacturer, Columbus Naval Iron Works, Columbus, Georgia.
- Boilers: Five horizontal return flues; 22 feet long by 4 feet in diameter; high pressure; manufacturer, Columbus Naval Iron Works, Columbus, Georgia.
- Propulsion system: One propeller, 10 feet 8 inches in diameter; originally designed for twin propellers; manufacturer, Columbus Naval Iron Works, Columbus, Georgia.
- Vessel disposition: Sold by US Navy, October 10, 1867, scrapped at later date.

Vessels Designed by William Graves

15. CSS *Charleston*

- Engines: One or two unknown direct acting; dimensions unknown; reportedly underpowered; manufacturer, probably J. M. Eason & Brother, Charleston, South Carolina.
- Boilers: Number and type unknown; dimensions unknown; probably low pressure; manufacturer, probably J. M. Eason & Brother, Charleston, South Carolina.
- Propulsion system: One propeller, 9 feet in diameter; manufacturer, probably J. M. Eason & Brother, Charleston, South Carolina.
- Vessel disposition: Wreck site buried in Charleston Harbor bottom, South Carolina, condition unknown.

16. CSS *Virginia II*

- Engines: Two single-cylinder horizontal or two British two-cylinder compound; approximately 40 inches in diameter by 38 inches stroke (if as shown on plans), or dimensions unknown (if British); referenced as having "nearly 800 horse[power]"; manufacturer, Tredegar Iron Works or Shockoe Foundry, Richmond, Virginia (if American), or unknown (if British).
- Boilers: Two horizontal firetubes, two furnaces each; 22 feet long by 9 feet high by 8 feet wide; probably low pressure; manufacturer, Tredegar Iron Works or Shockoe Foundry, Richmond, Virginia.
- Propulsion system: One propeller, 10 feet in diameter; manufacturer, Charlotte Navy Yard, Charlotte, North Carolina.
- Vessel disposition: Wreck site buried in James River bottom near Richmond, Virginia, condition unknown.

Large Sidewheelers

17. CSS *Nashville*

- Engines: Two single-cylinder inclined direct-acting; 30 or 32 inches in diameter by 9 feet stroke; high-pressure noncondensing, probably taken from the river steamboats *Capitol* or *Magenta*; manufacturer, Lent, South & Shipman, New Albany, Indiana (if the *Magenta*), or unknown (if the *Capitol*).
- Boilers: Seven horizontal return flues; 30 feet long by 40 or 42 inches in diameter; high pressure, probably taken from the river steamboats *Capitol* or *Magenta*; manufacturer, Lent, South & Shipman, New Albany, Indiana (if the *Magenta*), or unknown (if the *Capitol*).
- Propulsion system: Two sidewheels, 28 feet in diameter by 12 feet wide; independently turning; manufacturer, J. E. Montgomery and A. Anderson, Montgomery, Alabama.
- Vessel disposition: Sold by US Navy, November 22, 1867, scrapped at later date.

Diamond Hull Vessels

18. CSS *Tuscaloosa*

- Engines: Two adapted single-cylinder inclined direct-acting; dimensions unknown; high-pressure noncondensing, taken from the river steamboat *Chewala*; manufacturer unknown.
- Boilers: Number unknown, probably horizontal return flues; dimensions unknown; high pressure, taken from the river steamboat *Chewala*; manufacturer unknown.
- Propulsion system: Two propellers, approximately 8 feet in diameter,

geared to engines; manufacturer, William Penny & Company, Prattville, Alabama.

- Vessel disposition: Wreck site buried in Spanish River bottom near Mobile, Alabama, apparently nearly completely intact.

19. CSS *Huntsville*

- Engines: Two adapted single-cylinder inclined direct-acting; dimensions unknown, probably similar to the *Tuscaloosa*; high-pressure noncondensing, possibly taken from the river steamboat *John C. Calhoun*; manufacturer unknown.
- Boilers: Number unknown, probably horizontal return flues; dimensions unknown, probably similar to the *Tuscaloosa*; high pressure, possibly taken from the river steamboat *John C. Calhoun*; manufacturer unknown.
- Propulsion system: Two propellers, approximately 8 feet in diameter; geared to engines, completely ineffectual; manufacturer, Columbus Naval Iron Works, Columbus, Georgia.
- Vessel disposition: Wreck site buried in Spanish River bottom near Mobile, Alabama, apparently nearly completely intact.

20. CSS *Albemarle*

- Engines: Two single-cylinder horizontal direct-acting; 18 inches in diameter by 19 inches stroke; high-pressure noncondensing, may have been built from scratch; manufacturer uncertain, possibly Tredegar Iron Works, Richmond, Virginia.
- Boiler type: Two vertical watertubes, one furnace each; 15 feet 4 inches long by 5 feet 2 inches high by 4 feet 7 inches wide; high pressure, may have been built from scratch; manufacturer uncertain, possibly Tredegar Iron Works, Richmond, Virginia.
- Propulsion system: Two propellers, 6 feet in diameter; geared to engines; manufacturer, Charlotte Navy Yard, Charlotte, North Carolina.

21. CSS *Neuse*

- Engines: Two single-cylinder horizontal direct-acting; dimensions unknown, probably similar to the *Albemarle*; taken from Pugh's Mill, New Bern, North Carolina; manufacturer unknown.
- Boiler: One horizontal, probably return flue; 15 feet long, other dimensions unknown; taken from Pugh's Mill, New Bern, North Carolina; manufacturer unknown.
- Propulsion system: Two propellers, 6 feet in diameter; geared to engines; manufacturer unknown.
- Vessel disposition: Remaining hull and many artifacts on display at CSS *Neuse* Civil War Interpretive Center, Kinston, North Carolina.

22. CSS *Fredericksburg*

- Engines: Two single-cylinder horizontal direct-acting; 24 inches in diameter, other dimensions unknown; probably high-pressure noncondensing; manufacturer, Tredegar Iron Works or Shockoe Foundry, Richmond, Virginia.

- Boilers: Three horizontal, probably firetubes; approximately 15 feet long by 7 feet high by 5 feet in diameter; probably high pressure; manufacturer, Tredegar Iron Works or Shockoe Foundry, Richmond, Virginia.
- Propulsion system: Two propellers, 7 feet in diameter; unknown method of connection with engines; manufacturer, Tredegar Iron Works or Shockoe Foundry, Richmond, Virginia.
- Vessel disposition: Wreck site buried in James River bottom near Richmond, Virginia, condition unknown.

23. CSS *Missouri*

- Engines: Two single-cylinder inclined direct-acting; 24 inches in diameter by 7.5 feet stroke; high-pressure noncondensing, probably taken from the river steamboats *Grand Era* or *T. W. Roberts*; manufacturer unknown (if the *Grand Era*), or Lent, South & Shipman, New Albany, Indiana (if the *T. W. Roberts*).
- Boilers: Four horizontal return flues; 26 feet long by 40 inches in diameter; high-pressure, probably taken from the river steamboats *Grand Era* or *T. W. Roberts*; manufacturer unknown (if the *Grand Era*), or Lent, South & Shipman, New Albany, Indiana (if the *T.W. Roberts*).
- Propulsion system: One centerwheel, 22.5 feet indiameter by 17 feet wide; manufacturer probably Thomas Moore and John Smoker, Shreveport, Louisiana.
- Vessel disposition: Sold by US Navy, November 29, 1865; scrapped at later date.

Uncompleted Vessels (More Than 50 Percent Complete)

24. CSS *Mississippi*

- Engines: Three single-cylinder horizontal direct-acting; 30-inch bore by 36-inch stroke; high-pressure noncondensing; manufacturer, Patterson Iron Works (Jackson & Company), New Orleans, Louisiana.
- Boilers: Sixteen horizontal return flues; 30 feet long by 42 inches in diameter; high pressure, capable of approximately 1,500 HP; manufacturer, Patterson Iron Works (Jackson & Company), New Orleans, Louisiana.
- Propulsion system: Three propellers, 11 feet in diameter; one of the first triple-screw systems designed; manufacturer, Tredegar Iron Works, Richmond, Virginia (center shaft), Clark & Company, New Orleans, Louisiana (wing shafts).
- Vessel disposition: Burned remains broken up by US Navy.

25. CSS *Jackson*

- Engines (original design): Two single-cylinder inclined direct-acting; 30 inches in diameter by 9 feet stroke; high-pressure noncondensing; taken from the river steamboat *Time*; manufacturer unknown.
- Engines (final design): Two single-cylinder horizontal direct-acting; 36-inch bore by 36-inch stroke; high pressure noncondensing; manufacturer Columbus Naval Iron Works, Columbus, Georgia.
- Boilers: Four horizontal return flues; high pressure; taken from the river steamboat *Time*; retained in final design; dimensions unknown; manufacturer unknown.
- Propulsion system (original design): One centerwheel, approximately 20 feet in diameter by 20 feet wide; manufacturer, Columbus Naval Iron Works, Columbus, Georgia.
- Propulsion system (final design): Two propellers, 7.5 feet in diameter; independently turning; manufacturer, Columbus Naval Iron Works, Columbus, Georgia.
- Vessel disposition: Remaining hull and many artifacts on display at the Port Columbus Civil War Naval Museum, Columbus, Georgia.

26. CSS *Milledgeville*

- Engines: Two single-cylinder horizontal direct-acting; approximately 30 inches in diameter, other dimensions unknown; probably high pressure; manufacturer, Columbus Naval Iron Works, Columbus, Georgia.
- Boilers: Two horizontal; 14 feet long by 6 feet in diameter; probably high pressure; manufacturer, Columbus Naval Iron Works, Columbus, Georgia.
- Propulsion system: Two propellers, 7 feet in diameter; unknown method of connection with engines; manufacturer, Columbus Naval Iron Works, Columbus, Georgia.
- Vessel disposition: Wreck site in Savannah River near Savannah, Georgia; condition unknown.

27. CSS *Wilmington*

- Engines: Two single-cylinder horizontal direct-acting; 28-inch bore by 24-inch stroke; high-pressure nonondensing; manufacturer, Columbus Naval Iron Works, Columbus, Georgia.
- Boilers: Four horizontal, three furnaces each; approximately 16 feet long by 6.5 feet high by 8 feet wide; high pressure; manufacturer, Columbus Naval Iron Works, Columbus, Georgia.
- Propulsion system: Two propellers, 8 feet in diameter; geared to engines; manufacturer, Columbus Naval Iron Works, Columbus, Georgia.
- Vessel disposition: Burned on stocks, February 20, 1865.

Notes

Chapter 1. Origins, Background, and Technical Developments

1. James Phinney Baxter III, *The Introduction of the Ironclad Warship* (Cambridge, MA: Harvard University Press, 1933), 3.

2. William N. Still Jr., *Iron Afloat: The Story of the Confederate Armorclads* (Columbia: University of South Carolina Press, 1985), 5.

3. Prominent ironclads with non-American hulls and/or machinery were: CSS *Atlanta* (converted from the British blockade-runner *Fingal*), CSS *Stonewall* (a French-built ironclad corvette), and CSS *Texas* (an unfinished late-war model built in Richmond but utilizing British engines). Another ironclad not discussed in this work is USS *Eastport*, a large riverboat which the Confederates began armoring but the Union captured and completed. Its engines were, however, typical for a western river steamer.

4. P. Kean, *Report of Evidence Taken before a Joint Special Committee of Both Houses of the Confederate Congress to Investigate the Affairs of the Navy Department* (Richmond, VA: Geo. P. Evans, 1863), 108–17, 121–36; Still, *Iron Afloat*, 212; A. Robert Holcombe Jr., "The Evolution of Confederate Ironclad Design" (master's thesis, East Carolina University, Greenville, NC, 1993), 95–96, 109–12, 123–30.

5. Raimondo Luraghi, *A History of the Confederate Navy*, trans. Paolo E. Coletta (Annapolis, MD: US Naval Institute Press, 1996), 346.

6. Ibid., 345–49; William N. Still Jr., ed., *The Confederate Navy: The Ships, Men and Organization, 1861–65* (Annapolis, MD: US Naval Institute Press, 1997), 38–39.

7. Rodman L. Underwood, *Stephen Russell Mallory: A Biography of the Confederate Navy Secretary and United States Senator* (Jefferson, NC: McFarland, 2005), 5.

8. Joseph T. Durkin, *Confederate Navy Chief: Stephen R. Mallory* (Tuscaloosa: University of Alabama Press, 1954), 61.

9. Still, *Confederate Navy*, 2–3.

10. Ibid., 3.

11. Ibid., 10.

12. Ibid., 11. The Georgia navy also included the tugboats *Sampson* and *Resolute*.

13. Ibid.

14. Tom Henderson Wells, *The Confederate Navy: A Study in Organization* (Tuscaloosa: University of Alabama Press, 1971), 3–4.

15. Still, *Confederate Navy*, 14.

16. Durkin, *Confederate Navy Chief*, 156.

17. Holcombe, "Evolution," 7.

18. J. Thomas Scharf, *History of the Confederate States Navy from Its Organization to the Surrender of Its Last Vessel* (1886; repr., New York: Crown, 1977), 43.

19. Holcombe, "Evolution," 6–7.

20. Ibid., 22.

21. Ibid., 10.

22. Ibid., 39.

23. Ibid., 151–55. Estimates for the number of ironclads intended for construction, whether in contracts drawn up, materials collected, or simply viable proposals submitted to Mallory, range as high as seventy-five to one hundred vessels.

24. Ibid. Holcombe's thesis is the definitive work on the design of Confederate ironclads. It unfortunately remains unpublished.

25. Ibid., 83–84, 93–95.

26. Still, *Iron Afloat*, 80.

27. Holcombe, "Evolution," 44–45.

28. Panamerican Consultants and Tidewater Atlantic Research, *In Situ Archaeological Evaluation of the CSS Georgia, Savannah Harbor, Georgia: Final Report* (Savannah, GA: US Army Corps of Engineers, 2007), 14.

29. Holcombe, "Evolution," 68. Archaeological evidence suggests that at least one *Richmond* class vessel, the *Chicora*, was altered to carry an 8-foot screw. The size difference may reflect the fact that most of this vessel's machinery was removed from a tugboat that just happened to have an 8-foot propeller.

30. Ibid., 78.

31. Ibid., 83. No plans or drawings showing the *Nashville*'s hull below the waterline are known to exist. The ship's construction characteristics, its known draft, and the configuration of a scuttled and archaeologically examined sister vessel seem to suggest a standard hull.

32. Ibid., 92.

33. Ibid., 94. The twin-screw configuration is suggested by contemporary sources.

34. Ibid., 108.

35. Ibid., 125. Since virtually all Southern harbors and rivers were shallow, any decrease in draft that still allowed for an acceptable ordnance and armor configuration was a significant improvement.

36. Ibid., 128. In fact, the *Wilmington* had the greatest length-to-beam ratio of any Southern ironclad and should have handled quite well at sea.

37. Ibid., 147.

38. Kevin J. Foster, "Marine Steam Engines, 1807–1875: A Typology," in *Archaeology in Solution: Proceedings of the Seventeenth Annual Conference on Underwater Archaeology*, ed. John W. Foster and Sheli O. Smith (Salinas, CA: Fathom Eight, 1986), 67.

39. Ibid., 70; Andrew Lambert, ed., *Steam, Steel, and Shellfire: The Steam Warship, 1815–1905* (Edison, NJ: Conway Maritime Press, 1992), 172.

40. Foster, "Marine Steam Engines," 70.

41. Ibid.

42. Alan Bates, *The Western Rivers Engineroom Cyclopoedium* (Louisville, KY: Cyclopoedium Press, 1996), 7–8.

43. Lambert, *Steam, Steel, and Shellfire*, 173. "Additional components" in this case means all the extra connecting pieces and moving parts occasioned by the complex motion.

44. Foster, "Marine Steam Engines," 67.

45. Lewis Williams Taggart, "Technology and the Confederate Navy" (master's thesis, University of Tulsa, OK, 1991), 99.

46. Stephen Roper, *Handbook of Land and Marine Engines, including the Modeling, Construction, Running, and Management of Land and Marine Engines and Boilers*, 9th ed. (Philadelphia, PA: David McKay, 1897), 208.

47. Ibid.

48. Ibid., 352.

49. Bates, *Western Rivers Engineroom*, 10. Bates was referring to the Mississippi passenger boat *Eclipse* of post–Civil War steamboat racing fame. It is worth noting here that CSS *Mississippi*, which in many ways can be considered a hybrid of river and oceangoing steamboat designs, also carried sixteen boilers. Whether this huge ironclad would have been able to reach its intended top speed of 14 knots even with all that available power is debatable.

50. Charles Atherton, *On Marine Engine Construction and Classification* (London: John Weale Architectural Library, 1851), 40–41. Perfect combustion is an ideal process in which each particle of fuel receives the perfect amount of oxygen for complete combustion: a minimum number of by-products (i.e., ash and exhaust gases). Of course, in any real-world system such as a steam boiler, perfect combustion can only be approached, never reached.

51. Edgar C. Smith, *A Short History of Naval and Marine Engineering* (Cambridge, UK: Cambridge University Press, 1937), 133.

52. Ibid., 153.

53. Bates, *Western Rivers Engineroom*, 20.

54. Ibid., 16.

55. Ibid., 13.

56. Ibid., 15.

57. Wells, *Confederate Navy*, 107.

58. Ibid.

59. Ibid., 115.

60. Ibid., 117.

61. Ibid., 114.

62. David K. Brown, *Paddle Warships: The Earliest Steam-Powered Fighting Ships, 1815–1850* (London: Conway Maritime Press, 1993), 67.

63. Wells, *Confederate Navy*, 116.

64. William N. Still Jr., *Confederate Shipbuilding* (Athens: University of Georgia Press, 1987), 56–58. The navy had only forty-five railroad cars for carrying coal to Charleston, South Carolina; Savannah, Georgia; and Columbus, Georgia; and several other locations, such as Mobile, Alabama, had no rail connections to the mines at all.

65. Wells, *Confederate Navy*, 116. Smokeless coal was a very important wartime commodity, especially for vessels such as blockade-runners and warships. A lack of smoke by which to sight an incoming vessel greatly increased the chances of surprise against the enemy or in eluding pursuers.

66. Granville Bathe, *Ship of Destiny: A Record of the U.S. Steam Frigate Merrimac, 1855–1862* (St. Augustine, FL: n.p., 1951), 5–6.

67. Bates, *Western Rivers Engineroom*, 51.

68. Ibid., 10.

69. Bathe, *Ship of Destiny*, 6.

70. Ibid.

71. US Department of the Navy, *Official Records of the Union and Confederate Navies in the War of the Rebellion* (1894–1922; repr., Harrisburg, PA: National Historical Society, 1987), ser. 1, 10:792.

72. Wells, *Confederate Navy*, 149. The problem was so bad that it was common for about 20 percent of an ironclad's crew to be ill at any given time, and the crews of several vessels, such as the *Albemarle* and the *Baltic*, slept ashore or in the open air whenever possible.

Chapter 2. Conversions

1. Holcombe, "Evolution," 21.

2. "New Towboat," *Boston Evening Transcript*, December 5, 1854, 1; R. Thomas Campbell, "The First Confederate Ironclad," *America's Civil War*, July 2006, 48. The *Manassas* (ex–*Enoch Train*) was the first armored vessel completed in North America. A large ironclad, a pet project of wealthy industrialist and shipbuilder James B. Stevens, was begun in the 1840s in New York but never finished because of constant design changes and lack of funds. The Stevens battery was probably the first ironclad ordered for any navy. Durkin, *Confederate Navy Chief*, 63.

3. "Propeller Enoch Train, of Boston." *U.S. Nautical Magazine and Naval Journal* 3 (1855–1856): 221. The bow, at least, was noted for its strong construction, for it appears that the *Enoch Train* was also intended for some icebreaking and relief duties. Most sources describe the vessel as a tugboat *and* an icebreaker, or relief boat. Samuel Hartt Pook was a noted naval architect of clipper ships, and during the Civil War he designed the Union ironclad USS *Galena*. His father, Samuel Moore Pook, was also a naval architect and designed the seven City class ironclad gunboats used by the Union on the Mississippi River. They were commonly known as Pook Turtles.

4. "Towboat *Enoch Train*," *Boston Daily Journal*, March 13, 1855, 1; Holcombe, "Evolution," 21. The *Enoch Train* is also listed in contemporary newspaper accounts as being of 450 tons burthen. This was an old method of calculating the capacity of a ship based on its length and beam. Length between perpendiculars is the measurement taken from the aft end of the ship's stem to the forward end of the sternpost.

5. Paul H. Silverstone, *Civil War Navies, 1855–1883* (Annapolis, MD: US Naval Institute Press, 2001), 152.

6. "Biography of Harrison Loring," Harrison Loring Collection, Collection 226, Manuscript Collection Registers, Collections Research Center, Mystic Seaport, Mystic, CT, http://library.mysticseaport.org/manuscripts/coll/coll226.cfm.

7. J. Leander Bishop, *A History of American Manufacturers from 1608 to 1860* (Philadelphia, PA: Edward Young, 1864), 664–65, http://books.google.com/ebooks/reader?id=K1YdAAAAIAAJ.

8. *Atlantic Works v. Brady*, 107 US 192 (1883), http://openjurist.org/107/us/192/atlantio-works-v-brady-brady. The testimony of both parties offers an interesting glimpse into nineteenth-century dredging operations.

9. Ibid; "Towboat *Enoch Train*," 1.

10. James Brown to John H. Morrison, March 22, 1905, folder 49, box 1, Elwin Eldridge, Collection on John H. Morrison, 1881–1921, MS0006, Mariners' Museum Library, Christopher Newport University, Newport News, VA. Brown explicitly states, "One boiler, one smokestack, two (2) upright engines opposite each other."

11. Campbell, "First Confederate Ironclad," 48.

12. Maurice Melton, *The Confederate Ironclads* (New York: Thomas Yoseloff, 1968), 59.

13. Quoted in Still, *Iron Afloat*, 47.

14. Melton, *Confederate Ironclads*, 60.

15. Holcombe, "Evolution," 22. There remains much uncertainty about the *Manassas*'s actual appearance as converted or whether it was modified at various times in service. Several contemporary images and descriptions depict the ship with one center stack or two side by side, with different armor plate configurations, and with different numbers of hatches and even guns. An archaeological examination of the wreck could provide more definitive information on this historic vessel's actual design.

16. Ibid., 21–22.

17. "The Rebel War Vessels," *Boston Daily Journal*, August 9, 1861, 2. This article states, "The *Enoch Train* is nearly ready for sea. She is called a ram; [and] has had new and powerful boilers put into her."

18. Still, *Iron Afloat*, 48.

19. R. Thomas Campbell, *Midshipman in Gray: Selections from Recollections of a Rebel Reefer* (Shippensburg, PA: Burd Street Press, 1997), 55. James Morris Morgan (1845–1928) became a prominent naval figure during the war. He served at Charleston, South Carolina, and later on blockade-runners after the fall of New Orleans. His memoirs are a valuable firsthand account of the New Orleans campaigns.

20. John M. Stickney, *Promotion or the Bottom of the River: The Blue and Gray Naval Careers of Alexander F. Warley, South Carolinian* (Columbia: University of South Carolina Press, 2012), xi.

21. Ibid., xi, 1–2, 54–55, 60. USS *Mississippi* and its sister the *Missouri* were the first steam paddlewheel frigates built for the US Navy. The former became famous as a member of Commodore Matthew Perry's fleet during his ground-

breaking diplomatic mission to Japan, and during Warley's time aboard the ship was present during the British and French attack on the Chinese Taku forts (June 1859). The Second Opium War continued the efforts of the western European powers to expand their influence in Chinese domestic and foreign trade. Warley's service ashore was at the US Naval Observatory in Washington, DC.

22. Ibid., 64–65, 69, 74, 83, 89.

23. Ibid., 89. The retention of these three men was quite fortunate. Hardy was especially valuable because engineers of any experience were difficult to find, and pilot Mason had knowledge of the capricious Mississippi River currents and bars, which was especially invaluable in such an unwieldy vessel.

24. Campbell, "First Confederate Ironclad," 49.

25. Still, *Iron Afloat*, 48–50. USS *Richmond* was the last ship Warley served on while he was in the US Navy.

26. Stickney, *Promotion or the Bottom*, 90.

27. Ibid., 90–91. The *Manassas*'s crew was nearly gassed by the loss of draft from the destroyed uptakes. Engineer Hardy heroically ran topside and cut away the stays that were holding the collapsed stacks to the ship. With this loss of power and on one screw, the ironclad could barely move; hence Warley's decision to withdraw from the fight.

28. Still, *Iron Afloat*, 51; Stickney, *Promotion or the Bottom*, 93. USS *Richmond* was taken to Key West for the repair of grounding damage. Because of a misinterpretation of signal flags, USS *Vincennes*, which was also grounded during the affair, was abandoned and almost destroyed by its own crew before the error was discovered. Had Captain Hollins, the leader of the Confederate vessels, been more aggressive, he could have easily taken or destroyed the *Vincennes* and inflicted more damage on the other grounded ships. Such are the fortunes of war.

29. Stickney, *Promotion or the Bottom*, 93–94, 97, 101.

30. Robert U. Johnson and Clarence C. Buell, eds., *Battles and Leaders of the Civil War* (1887; repr., Harrisburg, PA: Archive Society, 1991), 2:90. This is one of very few statements that suggest that the *Manassas* had only one boiler. It is unknown why Chief Engineer Hardy was not present at this time.

31. Ibid., 91.

32. Campbell, "First Confederate Ironclad," 52.

33. Stickney, *Promotion or the Bottom*, 120. Warley was exchanged for Captain Alfred Gibbs of the Third US Cavalry. The chivalric practice of exchanging prisoners was prevalent during the early war years, before the continuing escalation of the conflict put a general stop to it.

34. Lambert, *Steam, Steel, and Shellfire*, 43. The British responded with the frigates *Mersey* and *Orlando*, which became the design antecedents of the magnificent HMS *Warrior*, Britain's first ironclad frigate. The *Mersey* and the *Orlando* were structurally weak, for their size pushed the limits of purely wooden construction. The *Warrior*, considered the world's first truly modern warship, has been beautifully restored and can be visited at Portsmouth Historic Dockyard, UK.

35. George F. Amadon, *Rise of the Ironclads* (Missoula, MT: Pictorial Histories, 1988), 9.

36. Carl D. Park, *Ironclad Down: The USS Merrimack–CSS Virginia from Construction to Destruction* (Annapolis, MD: US Naval Institute Press, 2007), 48. True oceangoing warships retained masts and sails far longer than coastal vessels did. In earlier years, engines often could not be totally relied upon; by the end of the century, sails were supplanted by reliable steam power as the prime mover.

37. Bathe, *Ship of Destiny*, 4.

38. "Archaeological Research at West Point Foundry Preserve," Scenic Hudson, http://web.archive.org/web/20070629101949/http://www.scenichudson.org/land_pres/wpfp_research.htm#history. Beginning in 1812, President James Madison recognized the foundry's future site as ideal for the production of modern artillery. Materials for manufacture were locally available, the location on the shore of the Hudson River allowed for efficient transportation, and the presence of the new US Military Academy directly across the river offered protection.

39. Bathe, *Ship of Destiny*, 4; Horace S. Hillery, "West Point Foundry," 1961, Rev. Horace S. Hillery Papers, 1953–1962, HC404, Historical Records Series, Historian's Collection, Putnam County Historian, Brewster, NY. The West Point Foundry, once one of the largest and most famous ironworks in the United States, went bankrupt in the early 1900s. The site is now composed of picturesque ruins surrounded by a preserve.

40. Park, *Ironclad Down*, 48. The manufacturer's guarantee was 8 knots under steam alone.

41. Amadon, *Rise of the Ironclads*, 9.

42. Holcombe, "Evolution," 21. The *Virginia*'s deep draft was arguably its greatest weakness.

43. Amadon, *Rise of the Ironclads*, 9.

44. Bathe, *Ship of Destiny*, 6. The tube plates, which contained hundreds of holes allowing water to circulate through the watertubes, were 0.5 inches thick.

45. Ibid., 4–6. Original draft copies of a set of large horizontal back-acting engines labeled as belonging to "Steamer Merrimack . . . West Point Foundry . . . October 26, 1854" can be found in Amadon, *Rise of the Ironclads*, 12–14. The engine arrangement shown there does not seem to match eyewitness accounts or other available data. I believe that either these drafts were mislabeled or the engine arrangement was reconfigured when installed to better fit inside the *Merrimack*'s deep narrow hull.

46. Bathe, *Ship of Destiny*, 7.

47. Amadon, *Rise of the Ironclads*, 16.

48. Park, *Ironclad Down*, 49–50.

49. Bathe, *Ship of Destiny*, 5.

50. Ibid. Benjamin Isherwood also helped establish and was appointed director of the US Navy's Bureau of Steam Engineering, just one of many notable accomplishments in his long career. His services during the Civil War in developing and testing new warships and machinery were indispensable to the Union cause. Isherwood is considered one of the US Navy's greatest engineers.

51. H. Ashton Ramsay, "Narrative of H. Ashton Ramsey, Chief Engineer, Confederate States Steamer Merrimack, during Her Engagement in Hampton Roads—1862," HA file, roll 13, frame 0353, Office of Naval Records and Library, Naval Records Collection, 1691–1945, Record Group 45, National Archives and Records Administration, College Park, MD (hereafter cited as ONRL, RG45); US Navy, *Official Records*, ser. 1, 7:758–59.

52. Johnson and Buell, *Battles and Leaders*, 1:716.

53. Ibid., 1:716–17.

54. Joseph A. Gutierrez Jr., "Confederate Naval Ordnance" (master's thesis, East Carolina University, Greenville, NC, 1977), 104–15.

55. Alan B. Flanders, *John L. Porter: Naval Constructor of Destiny* (White Stone, VA: Brandylane, 2000), ix–xi.

56. Ibid., 7–9, 21–22, 27, 37. Porter was appointed a US naval constructor in 1859. His 1846 ironclad plans became the basis for the *Richmond* class ironclads when he sided with the Confederacy. Porter's work on the steam frigate USS *Colorado* also proved fateful: his familiarity with that ship's design allowed him to effectively supervise the deconstruction of the burned *Merrimack* and use its remaining structure as the basis for CSS *Virginia*.

57. Ibid., ix, 38–39. Porter was very successful in the US Navy; it is not difficult to imagine him as the chief naval constructor had he sided with the Union. The very fact that he was appointed to supervise the complete rebuilding of the venerable 1797 frigate USS *Constellation* into virtually a new ship in 1853–1854 speaks of the high esteem in which he was held and the skills he had developed. Loyalty to one's home state over that to the nation was more common in those days of states' rights and a weaker federal government.

58. Ibid., 38.

59. Holcombe, "Evolution," 16. The late appointment to chief naval constructor finally gave Porter a rank equal to the duties he had performed since 1861 in designing the Confederacy's warships. Postwar, Porter worked in several shipyards and ironworks before becoming superintendent of the Norfolk County, Virginia, ferries. He died in 1893.

60. George M. Brooke, Jr., ed., *Ironclads and Big Guns of the Confederacy: The Journal and Letters of John M. Brooke* (Columbia: University of South Carolina Press, 2002), xiii–xiv.

61. Ibid., 1–2.

62. Ibid., 14.

63. Ibid., 22.

64. Ibid., 28. Brooke determined that at least two layers of 2-inch-thick plating (the maximum thickness that could be rolled by local ironworks) was optimal. Combined with an approximately 35-degree angle of deflection in the construction of the casemate and a wood backing 2 feet thick, this layout was used by virtually every Confederate ironclad afterward.

65. Gutierrez, "Confederate Naval Ordnance," 90, 93–94, 100.

66. Brooke, *Ironclads and Big Guns*, 110, 204–5. This was the high point of Brooke's career. After the war he applied for and received a presidential pardon.

Thereafter he lived a fairly quiet life teaching astronomy and related subjects at the Virginia Military Institute until his retirement to professor emeritus in 1899. Brooke died in 1906.

67. Gutierrez, "Confederate Naval Ordnance," 118.

68. Flanders, *John L. Porter,* 50; Robert J. Driver Jr., *Confederate Sailors, Marines, and Signalmen from Virginia and Maryland* (Westminster, MD: Heritage Books, 2007), 369.

69. Flanders, *John L. Porter,* 50. At the war's end Williamson was paroled at Greensboro, North Carolina, and took the oath of allegiance to the United States at Richmond on May 15, 1865. He seems to have been worn out, and he died at Norfolk in October 1870. Driver, *Confederate Sailors,* 369.

70. Still, *Confederate Shipbuilding,* 27.

71. Flanders, *John L. Porter,* 76.

72. Johnson and Buell, *Battles and Leaders,* 1:694. John Taylor Wood gained fame for gallantry in action as the commander of the *Virginia*'s aft pivot gun during the fight with the *Monitor* and later at Drewry's Bluff on the James River in Virginia. His war service afterward was busy and daring: successful expeditions to capture the federal warships *Satellite, Reliance,* and *Underwriter*; an attempt to free prisoners of war from Johnson's Island prison in Ohio; an appointment to President Jefferson Davis's staff; and the capture of Portsmouth, North Carolina. Wood's wartime career culminated in his appointment as the captain of the commerce raider CSS *Tallahassee* from 1864 to 1865. At the war's end he escaped to Cuba, then lived in Nova Scotia until his death in 1904. Driver, *Confederate Sailors,* 373–74.

73. Ramsay, "Narrative," frame 0351; Driver, *Confederate Sailors,* 276.

74. Ibid., frame 0353.

75. Ibid.

76. Ibid., frames 0353–0354.

77. Ibid., frame 0355.

78. Ibid.

79. Ibid., frame 0359.

80. Ibid., frame 0360.

81. Ibid.

82. US Navy, *Official Records,* ser. 1, 7:758–59.

83. Ibid., 759.

84. Alan B. Flanders and Neale O. Westfall, eds., *Memoirs of E. A. Jack: Steam Engineer, CSS Virginia* (White Stone, VA: Brandylane, 1998), 7.

85. Ibid., 13. E. V. White's station at the bell pulls show that the *Merrimack,* or *Virginia,* was fitted with typical engine room bell signals in those pre-engine order telegraph days.

86. Ibid., 10.

87. Ibid., 15.

88. Ibid., 16.

89. Ibid., 17.

90. Ibid., 21. The ironclad was soon christened CSS *Arkansas.*

91. Holcombe, "Evolution," 25.

92. "The Tow-Boat 'Baltic,'" *New Albany (IN) Tribune*, December 10, 1856, 3. New Albany was one of the country's largest shipbuilding centers for the river steamboat industry before the war.

93. Mart Tellon, "In the Heyday of Steamboating in New Albany," *New Albany (IN) Weekly Ledger*, May 28, 1924, 2, http://webcache.googleusercontent.com/search?q=cache:http://www.nafclibrary.org/wp-content/uploads/2013/08/HeydayofSteamboating.pdf.

94. "Tow-Boat 'Baltic,'" 3.

95. US Navy, *Official Records*, ser. 1, 19:198–200.

96. "Steamer Baltic," *New Albany (IN) Daily Ledger*, December 16, 1856, 3; "Our Boat Building . . . ," *Louisville (KY) Daily Courier*, January 19, 1857, 3.

97. "Tow-Boat 'Baltic,'" 3; Still, *Iron Afloat*, 80.

98. Clyde E. Wilson, Gunboat *Baltic* Papers, 1938, folder 10, 2–3, Public Information Subject Files: Navy, SG11146, Alabama Department of Archives and History, Montgomery. The *Baltic*'s speedy conversion was not speedy enough for Flag Officer Victor Randolph, who commanded the Mobile station. He accused the committee of pandering to the mechanics and generally taking too long in the conversion process. Still, *Iron Afloat*, 80.

99. US Navy, *Official Records*, ser. 2, 1:248.

100. Ibid., ser. 1, 22:226.

101. Still, *Iron Afloat*, 201. More information on Johnston and the *Tennessee* is presented in chapter 5.

102. US Navy, *Official Records*, ser. 1, 21:886.

103. Driver, *Confederate Sailors*, 305.

104. US Navy, *Official Records*, ser. 1, 18:660.

105. Ibid., ser. 1, 22:226–27.

106. "Tow-Boat 'Baltic,'" 3; US Navy, *Official Records*, ser. 1, 22:226. Slight differences in measurement are common to many steam plants of this period. The difference in boiler width could reflect a measurement of the inside versus outside diameters. The 4-foot difference in length may be due to the newspaper accounts being contemporary with the machinery's manufacture—that is, their measurements may reflect the dimensions of the bare boiler shell, minus its insulation and supporting framework.

107. US Navy, *Official Records*, ser. 1, 22:226.

108. Ibid.

109. James L. Mooney, ed., *Dictionary of American Naval Fighting Ships* (Washington, DC: US Department of the Navy, 1963), 2:502. The entry for the *Baltic* states that it was "sold by the US Government on 31 December 1865." Therefore it was almost certainly broken up sometime in the new year. The vessel's sinking condition and lack of further records seem to preclude any continued service under another guise.

Chapter 3. Early Nonstandard Designs

1. Holcombe, "Evolution," 6; Melton, *Confederate Ironclads*, 63.

2. James M. Merrill, "Confederate Shipbuilding at New Orleans," *Journal of Southern History* 28, no. 1 (1962): 87.

3. Holcombe, "Evolution," 47. Whether the casemate as finished was actually angled at 45 degrees is uncertain. A primary reason for the fewer number of guns was a delay in getting them to New Orleans.

4. Still, *Confederate Navy*, 53, 97.

5. Frederick Way Jr., *Way's Packet Directory, 1848–1994* (Athens: Ohio University Press, 1983), 224.

6. "Ingomar," *Louisville (KY) Daily Courier*, January 8, 1855, 3.

7. Ibid.; Kean, *Report of Evidence*, 378; Way, *Packet Directory*, 224.

8. Way, *Packet Directory*, n.p. Local works of history not available elsewhere, and local newspapers are probably the best way to locate such information, if it still exists.

9. John E. Kleber, ed., *The Encyclopedia of Louisville* (Lexington: University Press of Kentucky, 2001), 422.

10. Holcombe, "Evolution," 48; Campbell, *Midshipman in Gray*, 57; William Ragan, "The *Louisiana*" (unpublished manuscript, 2012), 1. Ragan, who has extensively researched the New Orleans ironclads, helpfully pointed out a primary source quote that the reduction of the fore wheel would result in an increased flow to the aft wheel. As with the other features of the *Louisiana*'s propulsion, this idea makes sense on paper but was a failure in practice.

11. Kean, *Report of Evidence*, 378.

12. Ibid., 372.

13. Bishop, *History*, 68–69.

14. Ulysses Grant Duffield, *The Neafie & Levy Ship and Engine Building Company (Penn Works)* (Philadelphia, PA: Franklin, 1896), 5–6.

15. Edwin T. Freedley, ed., *United States Mercantile Guide: Leading Pursuits and Leading Men, a Treatise on the Principal Trades and Manufactures of the United States* (Philadelphia, PA: Edward Young, 1856), 295, http://books.google.com/ebooks/reader?id=qvcYAAAAYAAJ.

16. Campbell, *Midshipman in Gray*, 57.

17. Reanie, Neafie & Company Papers, 1844–1861, folder 4, 145, 196, MS102, Mariners' Museum Library, Christopher Newport University, Newport News, VA; Work Projects Administration, *Ship Registers and Enrollments of New Orleans, Louisiana*, vol. 5, *1851–1860* (Baton Rouge: Louisiana State University, 1942), 12, 63, http://www.youngsanders.org/shipvol5.pdf.

18. Ragan, "*Louisiana*," 7.

19. Reanie, Neafie & Company Papers, folder 4, 145, 196.

20. Work Projects Administration, *Ship Registers*, 5:12, 5:63. http://www.youngsanders.org/shipvol5.pdf.

21. Kean, *Report of Evidence*, 372, 378.

22. Ragan, "*Louisiana*," 1.

23. Kean, *Report of Evidence*, 378. It is probable that Kirk's establishment suffered from a lack of workmen and/or materials to speedily prosecute the work—a common problem encountered by Southern industry during the war.

24. Holcombe, "Evolution," 47.

25. Driver, *Confederate Sailors*, 234–35.

26. Campbell, *Midshipman in Gray*, 57.

27. Kean, *Report of Evidence*, 29.

28. Johnson and Buell, *Battles and Leaders*, 2:48. Whittle, though young, was already distinguished for his service on the commerce raider CSS *Nashville*. He later served on the gunboat CSS *Chattahoochee* after his exchange from Fort Warren following his capture at New Orleans. Whittle achieved fame at the end of his career in the Confederate navy as the first lieutenant of the legendary commerce raider and last Confederate warship CSS *Shenandoah*. He was a prominent leader in several business interests postwar before his death in 1920. Driver, *Confederate Sailors*, 363.

29. Ramsay, "Narrative," frame 0354; Driver, *Confederate Sailors*, 212.

30. Driver, *Confederate Sailors*, 212.

31. Ramsay, "Narrative," frames 0353–0354.

32. Holcombe, "Evolution," 49.

33. US Navy, *Official Records*, ser. 1, 18:318. Youngblood was born in South Carolina in 1830. He was the senior engineer on the *Louisiana* but was technically ranked as a second assistant engineer until his promotion to first assistant, then chief, engineer later in the war. After his exchange from prison, Youngblood served at Jackson, Mississippi, and later on the gunboat CSS *Morgan* at Mobile until the war's end. After the war he lived in New Orleans until his death in 1914. "Y-Z, C.S. Navy Sailors," *NC Civil War Sailors Project: Identifying NC Sailors and Marines*, http://rblong.net/sailor/y-z.html.

34. Holcombe, "Evolution," 39.

35. Still, *Iron Afloat*, 62.

36. Myron J. Smith Jr., *The CSS Arkansas: A Civil War Ironclad on Western Waters* (Jefferson, NC: McFarland, 2011), 31, 34. Brown was promoted to commander after assuming his post on the *Arkansas*. After the war he lived a quiet life as a farmer in Mississippi and Texas. He died in 1889. A close friend from Brown's prewar days in the US Navy, Commander Henry Walke, described him as "one of the best Confederate officers." During the war Brown and Walke actually met in combat: Brown commanding the *Arkansas*, and Walke commanding USS *Carondelet*.

37. Still, *Iron Afloat*, 64.

38. Johnson and Buell, *Battles and Leaders*, 3:572.

39. Ibid.

40. Smith, *CSS Arkansas*, 45.

41. "Leeds Iron Foundry—New Orleans, LA," Waymarking, http://www.waymarking.com/waymarks/WM14QA_Leeds_Iron_Foundry_New_Orleans_LA.

42. Holcombe, "Evolution," 39–40; Flanders and Westfall, *Memoirs of E. A. Jack*, 26. The boiler dimensions are unknown. Multiple crew-member testimonies state that the *Arkansas* burned coal. This is noteworthy not only because of the general scarcity of coal for Confederate vessels but also because the *Arkan-*

sas was built on the Mississippi River system, where virtually every boat burned wood fuel.

43. Holcombe, "Evolution," 40.

44. Johnson and Buell, *Battles and Leaders*, 3:572.

45. Flanders and Westfall, *Memoirs of E. A. Jack*, 23.

46. Johnson and Buell, *Battles and Leaders*, 3:573.

47. Ibid., 575.

48. Flanders and Westfall, *Memoirs of E. A. Jack*, 23.

49. Johnson and Buell, *Battles and Leaders*, 3:575.

50. "Biographical Information," George W. Gift Papers, 1862–1876, Collection 01152, Southern Historical Collection, University of North Carolina, Chapel Hill, http://www2.lib.unc.edu/mss/inv/g/Gift,George_W.html; Smith, *CSS Arkansas*, 117.

51. H. Allen Gosnell, *Guns on the Western Waters: The Story of River Gunboats in the Civil War* (Baton Rouge: Louisiana State University Press, 1949), 121.

52. Ibid., 125.

53. Still, *Iron Afloat*, 75.

54. Smith, *CSS Arkansas*, 71.

55. Gosnell, *Guns on the Western Waters*, 132.

56. Driver, *Confederate Sailors*, 60–61. Unfortunately, City apparently left no memoirs.

57. Flanders and Westfall, *Memoirs of E. A. Jack*, 25.

58. Ibid., 25–26.

59. Ibid.

60. Gosnell, *Guns on the Western Waters*, 132.

61. Flanders and Westfall, *Memoirs of E. A. Jack*, 26. 62. Gosnell, *Guns on the Western Waters*, 134.

63. Still, *Iron Afloat*, 76.

64. Gosnell, *Guns on the Western Waters*, 134–35.

65. Flanders and Westfall, *Memoirs of E. A. Jack*, 26–27.

66. Gosnell, *Guns on the Western Waters*, 135.

67. Smith, CSS *Arkansas*, 108–9.

68. Holcombe, "Evolution," 49–50. The hull is variously speculated to be based on such wide-ranging designs as those of CSS *Mississippi* and CSS *Louisiana*, which were quite bargelike with few or no curves; to CSS *Albemarle* and CSS *Neuse*, which had curves only around the bow and propellers; and to the *Richmond* class vessels, which had shiplike hulls. It is probably safe to assume that the *Georgia*'s hull was most similar to those first listed, possibly being based on the hull design of Constructor Porter's June 1861 harbor defense ironclad plans.

69. Panamerican and Tidewater, *In Situ Archaeological Evaluation*, 65. Much remains of the casemate, machinery, and other artifacts.

70. Holcombe, "Evolution," 50.

71. Mark Swanson and Robert Holcombe, *CSS Georgia: Archival Study* (Stone Mountain, GA: New South Associates, 2007), 45–48. One of the vessels Miller helped repair while apprenticed to a New York shipbuilder in the 1820s was the

British paddle steamer *Aaron Manby*, the first iron-hulled vessel to make a sea voyage. He also submitted an early proposal for an ironclad to the Confederate Navy Department in late 1861. Nothing was done with it, but how closely the *Georgia* may have resembled that proposal is unknown.

72. Panamerican and Tidewater, *In Situ Archaeological Evaluation*, 8.

73. Holcombe, "Evolution," 50.

74. Swanson and Holcombe, *CSS Georgia*, 33–34.

75. Panamerican and Tidewater, *In Situ Archaeological Evaluation*, 14.

76. Still, *Iron Afloat*, 87. The use of green wood in new hulls was an almost universal problem resulting from hasty wartime construction and the lack of seasoned wood.

77. Swanson and Holcombe, *CSS Georgia*, 66.

78. Panamerican and Tidewater, *In Situ Archaeological Evaluation*, 24, 64.

79. William N. Still Jr., *The Savannah Squadron* (Savannah, GA: Coastal Heritage Press, 1989), 9.

80. Swanson and Holcombe, *CSS Georgia*, 67; Panamerican and Tidewater, *In Situ Archaeological Evaluation*, 64.

81. Swanson and Holcombe, *CSS Georgia*, 78–79. Gwathmey commanded the armed steamer CSS *Resolute* (formerly a tugboat) before his appointment to the *Georgia*.

82. Ibid., 77, 86–87.

83. Ibid., 90. Why the vessel was sunk instead of destroyed is uncertain. According to Swanson and Holcombe, the "*Georgia* may have been scuttled simply because that was the only option left. How the scuttling was achieved is not known. In a vessel that leaked as badly as *Georgia*, it might not have taken much to sink the ship."

84. Ibid., 40.

85. Panamerican and Tidewater, *In Situ Archaeological Evaluation*, 42–44.

86. Ibid., 59.

87. Ibid., 61.

Chapter 4. The Richmond Class Ironclads

1. A. Robert Holcombe Jr., "The *Richmond* Class Confederate Ironclads," *CHAB (Confederate Historical Association of Belgium) News* 6 (1978): 16. The Confederacy did in fact turn to European powers (Britain and France) to construct its seagoing ironclads, but these efforts were unsuccessful. Only one ironclad, the French-built CSS *Stonewall*, was ever delivered into Confederate hands. It arrived in American waters in May 1865, too late to take part in the war.

2. Ibid.

3. Ibid.

4. Holcombe, "Evolution," 63–64. The 1846 plan called for propulsion by the submerged Hunter wheel. This device, named for Lieutenant William W. Hunter, was an attempt to combine the low underwater profile and high effi-

ciency of the propeller with the paddlewheel concept. The wheel was mounted horizontally above the keel. Several different vessels were used for trials; the last was USS *Allegheny,* which Porter was working on when he conceived his first ironclad. All the trials confirmed the great inefficiency and difficulty of maintenance of the Hunter wheel. By that time the screw propeller was coming into widespread use, and the idea was dropped.

5. Still, *Iron Afloat,* 94.

6. Ibid.

7. Holcombe, "*Richmond* Class," 17.

8. Erik Heyl, *Early American Steamers* (Buffalo, NY: Erik Heyl, 1953), 14.

9. US Light House Establishment, *Manual of Light-House Engineering, Embracing Plans of Towers, Light Vessels, Beacons, Bouys, and Other Aids to Navigation, Designed and Executed under the Direction of the Light House Board, and Exhibiting the Present System of Lenses, Lamps, Implements* (Washington, DC: Light-House Board, 1869), plate 52.

10. Ibid., plates 52–53. It is remarkable to observe how the general form of US lightships changed so little over the years.

11. A. Robert Holcombe Jr., e-mail to author, April 17, 2012.

12. Edwin L. Combs III, "On Duty at Wilmington: The Confederate Navy on the Cape Fear River" (master's thesis, East Carolina University, Greenville, NC, 1996), 3. The *Arctic*'s configuration as an ironclad floating battery is unknown. The term *floating battery* suggests that no new propelling machinery was installed. The vessel was scuttled by the Confederates after the fall of Fort Fisher.

13. Holcombe, "Evolution," 16.

14. Martin D. Peebles, "CSS *Raleigh*: The History and Archaeology of a Civil War Ironclad in the Cape Fear River" (master's thesis, East Carolina University, Greenville, NC, 1996), 10.

15. US Navy, *Official Records,* ser. 1, 8:68.

16. Charles B. Dew, *Ironmaker to the Confederacy: Joseph R. Anderson and the Tredegar Iron Works* (Wilmington, NC: Broadfoot, 1987), 3–5.

17. Ibid., 2.

18. Ibid., 265.

19. Kathleen Bruce, *Virginia Iron Manufacture in the Slave Era* (New York: Century Company, 1931), 296.

20. John M. Coski, *Capital Navy: The Men, Ships and Operations of the James River Squadron* (New York: Savas Beatie, 2005), 72.

21. Bruce, *Virginia Iron Manufacture,* 296–97.

22. Coski, *Capital Navy,* 282–83.

23. Still, *Confederate Navy,* 68.

24. US Navy, *Official Records,* ser. 2, 2:256. Five knots was actually about the average speed for most Southern ironclads.

25. Coski, *Capital Navy,* 80. The use of *engine* in the singular in one account and plural in the next, describing the same vessel, has led to much confusion over the years. This is true of several other vessels besides the *Richmond.*

26. Ibid., 91. The usual problems of labor and materials scarcity were the primary cause.

27. US Navy, *Official Records,* ser. 1, 8:841–42. Lieutenant John Brooke noted some specific details of the *Richmond*'s engine in operation: 45 RPM were obtained at a maximum boiler pressure of 28 PSI. Brooke, *Ironclads and Big Guns,* 118.

28. Coski, *Capital Navy,* 91.

29. Driver, *Confederate Sailors,* 376.

30. Henry X. Wright, report to Robert B. Pegram, January 28, 1863, folder 13, Henry X. Wright Papers, 1859–1874, MS0072, Mariners' Museum Library, Christopher Newport University, Newport News, VA; Wright, General Order, July 29, 1863, folder 14, Wright Papers.

31. US Navy, *Official Records,* ser. 1, 10:792.

32. Coski, *Capital Navy,* 155–56. The loss of New Orleans remained a major blemish on Mitchell's service record despite the fact that the investigation of the Navy Department cleared him of any negligence or wrongdoing. He therefore must have been that much more determined to properly prepare his new squadron for battle.

33. Ibid., 197–207. The main purpose of the Confederate offensive was to reach and destroy General Ulysses S. Grant's supply depot downriver at City Point, which was vital to the continuing Union siege of Petersburg.

34. Still, *Iron Afloat,* 115.

35. Ernest M. Lander Jr., "Charleston: Manufacturing Center of the Old South," *Journal of Southern History* 26, no. 3 (1960): 333.

36. Ibid., 333–34. The Best Friend of Charleston was actually built at the West Point Foundry in New York State and disassembled for shipment to the South. It became famous, originally as the first steam locomotive built entirely in the United States, and later for a fatal boiler explosion (another unfortunate first for steam locomotion in the United States). The design was unusual, with an upright boiler positioned behind the engineer's position up forward between the cylinders.

37. Ibid., 334.

38. Reanie, Neafie & Company Papers, folder 4, 94.

39. Ibid.; William M. Lytle and Forrest R. Holdcamper, *Merchant Steam Vessels of the United States 1790–1868* (Staten Island, NY: Steamship Historical Society of America, 1975), 4.

40. Reanie, Neafie & Company Papers, folder 4, 94; R. Thomas Campbell, *Engineer in Gray: Memoirs of Chief Engineer James H. Tomb, CSN* (Jefferson, NC: McFarland, 2005), 54–55.

41. Stickney, *Promotion or the Bottom,* 121, 137–42.

42. Campbell, *Engineer in Gray,* 11, 25, 42, 65, 101–8. There were ultimately only twelve chief engineers in the Confederate navy. Tomb was later an advisor for Brazilian naval forces during the War of the Triple Alliance (1864–1870). He died in 1929.

43. Ibid., 54–55.

44. Ibid., 58–59. This fear was well-founded: when CSS *Virginia* rammed USS *Cumberland* at Hampton Roads and CSS *Albemarle* rammed USS *Southfield* at Plymouth, both Confederate vessels had difficulty backing out and were nearly pulled down with their sinking targets.

45. Charles James Wexler, "'The Ordinary Gunboats': The CSS *Chicora*, the CSS *Palmetto State*, and the Battle off Charleston Harbor, January 31, 1863" (master's thesis, Citadel, Charleston, SC, 2008), 72–75.

46. Still, *Iron Afloat*, 115.

47. US Navy, *Official Records*, ser. 1, 15:230. A boiler, simply because of temperature and chemical stresses arising from its function, wore out fairly quickly even when well maintained. In contrast, a vessel's engine could last for decades. Therefore, even though the *Chicora*'s boilers were new in 1862, by the last year of the Civil War they were ready for replacement. A major contributing factor was a probable lack of regular maintenance, a result of the exigencies of war. As long as the boilers worked at some level (or did not explode), the *Chicora* could continue its crucial defense duties.

48. Lynn Harris, "Wreck of the Golden Spike," *Legacy* 8, no. 1 (2003): 9.

49. Wexler, "Ordinary Gunboats," 43.

50. Holcombe, "*Richmond* Class," 18; Stickney, *Promotion or the Bottom*, 126. The *Palmetto State*, like the CSS *Georgia*, was often known as the Ladies' Gunboat.

51. Holcombe, "Evolution," 67. A more efficient disposition of the gunports may have been the attempted purpose of the octagonal casemate, whereas positioning the pilothouse aft may have been for ease of communication with the engine room directly below. The *Palmetto State*'s sister the *Chicora* also had an aft pilothouse as well as one forward, and their sister the *Savannah* had a pilot hatch placed abaft the stack in a similar manner to that of the *Palmetto State*. Forward visibility was not necessarily a problem, because of the low speeds involved, the extensive use of local pilots, and the stationing of lookouts on the top deck. These points are explained in greater detail in Peebles, "CSS *Raleigh*," 11–13.

52. Wexler, "Ordinary Gunboats," 39.

53. Lander, "Charleston," 332.

54. Wexler, "Ordinary Gunboats" 14.

55. Reanie, Neafie & Company Papers, folder 5, 211.

56. Ibid.

57. William H. Parker, *Recollections of a Naval Officer, 1841–1865* (New York: Charles Scribner's Sons, 1883), 288–289.

58. Driver, *Confederate Sailors*, 256. Parker was just as successful after the war, when he first worked for, and then became the president of, the Pacific Mail Steamship Company. He later was the president of Maryland Agricultural College. He next served as a US consul. He died in Washington, DC, in 1896.

59. Wexler, "Ordinary Gunboats," 71–72.

60. US Navy, *Official Records*, ser. 1, 15:678.

61. Holcombe, "*Richmond* Class," 17.

62. Gordon P. Watts, *Underwater Archaeological Documentation of the*

Remains of the Ironclad CSS North Carolina, Brunswick County, North Carolina (Wilmington, NC: US Army Engineer District, 1999), 13.

63. Combs, "On Duty at Wilmington," 61. The Cape Fear area produced the majority of naval stores (e.g., pitch and turpentine) in use on US vessels before the Civil War.

64. Ibid., 62.

65. Ibid., 1–2.

66. "The Buffalo Dry Dock," *The S.S. Canadiana History Book*, SS Canadiana, http://sscanadiana.com/buffdrydock.html.

67. Mooney, *Dictionary*, 2:577.

68. Watts, *Underwater Archaeological Documentation*, 13.

69. Ibid., 9.

70. Driver, *Confederate Sailors*, 260. After the war Peek was a machinist and a merchant in Hampton, Virginia. He died in 1887.

71. Charles Smith Peek to "Sis," July 1, 1863, letter 4, Charles Smith Peek Papers, 1863–1865, Collection of Dr. Charles Peery, Charleston, SC.

72. Peek to "Sis," July 30, 1863, letter 7, Peek Papers.

73. Peek to "Sis," August 12, 1863, letter 8, Peek Papers.

74. Peek to "Sis," July 6, 1864, letter 31, Peek Papers.

75. Peek to "Sister," July 8, 1864, letter 32, Peek Papers.

76. Peek to "Bro," July 24, 1864, letter 33, Peek Papers.

77. Watts, *Underwater Archaeological Documentation*, 14. Smithville was renamed Southport in the 1880s. The *North Carolina*'s wreck site is located next to Battery Island, just across from the picturesque town waterfront and the North Carolina Maritime Museum at Southport.

78. Ibid., 14.

79. Peebles, "CSS *Raleigh*," 35, 98.

80. Ibid., 35, 99. The plate, if it still exists, is probably obscured by marine growth, debris, or concretion. Finding such an object underwater is always difficult even in good conditions.

81. William N. Still Jr., "Facilities for the Construction of War Vessels in the Confederacy," *Journal of Southern History* 31, no. 3 (1965): 294.

82. US Navy, *Official Records*, ser. 1, 8:89. Robins is listed as a ship's carpenter. His account should be seriously considered because the details he lists about the *Raleigh* and the *North Carolina* are supported by other sources and archaeological evidence. Many other contrabands and army deserters reporting to Union officials were prone to exaggeration or hearsay in their accounts.

83. M. Smith to Gilbert Elliott, November 4, 1861, folder 36, William Francis Martin Papers, 1787–1884, Southern Historical Collection, University of North Carolina, Chapel Hill. The dimensions given for these boats reveal that they were of the generic Confederate wooden gunboat design intended for mass production at the beginning of the war. Ultimately, only a few of these vessels were ever completed. The most famous is CSS *Chattahoochee*, the stern of which is preserved today in Columbus, Georgia.

84. Confederate States Navy Department, invitation for machinery construction bids and contract with George Mooney, October 3, 1861, entry 59, US Department of the Treasury, Confederate Records, 1833–1878, Record Group 365, National Archives and Records Administration, College Park, MD.

85. A. Robert Holcombe Jr., e-mail to author, July 27, 2011.

86. Combs, "On Duty at Wilmington," 63.

87. Peebles, "CSS *Raleigh*," 35, 39–42.

88. Driver, *Confederate Sailors*, 178–79, 244. Jones was also a Mexican War veteran. After the Civil War he served in the Argentine navy and surveyed the Rio de la Plata before settling in California, where he died in 1910.

89. Peebles, "CSS *Raleigh*," 45–47.

90. Still, *Iron Afloat*, 166–67.

91. Peebles, "CSS *Raleigh*," 48.

92. James Sprunt, *Chronicles of the Cape Fear River, 1660–1916* (Raleigh, NC: Edwards and Broughton, 1916), 482.

93. Peebles, "CSS *Raleigh*," 63, 68.

94. Ibid., 78–103.

95. Ibid., 98–99.

96. S. R. Mallory to James H. Warner, April 10, 1862, Area 8 file, frame 0258, ONRL, RG45.

97. Maxine Turner, *Navy Gray: A Story of the Confederate Navy on the Chattahoochee and Apalachicola Rivers* (Tuscaloosa: University of Alabama Press, 1988), 9.

98. Diffie W. Standard, *Columbus, Georgia, in the Confederacy* (New York: William-Frederick Press, 1954), 40.

99. Ibid., 40–41.

100. Turner, *Navy Gray*, 149, 154.

101. Driver, *Confederate Sailors*, 212–13. Almost nothing is known about McLaughlin's postwar career. He was listed as a member of the Army & Navy Society, and he died in Baltimore in 1884.

102. Ibid., 352–53.

103. Turner, *Navy Gray*, 236.

104. Still, *Savannah Squadron*, 9.

105. Turner, *Navy Gray*, 261–64.

106. Holcombe, "*Richmond* Class," 19.

107. S. R. Mallory, orders to W. W. Hunter, July 18, 1863, folder 10, box 1, Savannah Squadron Papers, 1862–1865, Manuscript, Archives, and Rare Book Library, Robert W. Woodruff Library, Emory University, Atlanta, GA.

108. Driver, *Confederate Sailors*, 166.

109. Engineering report, July 18, 1863, folder 10, box 1, Savannah Squadron Papers.

110. Quoted in Still, *Savannah Squadron*, 16.

111. Turner, *Navy Gray*, 165. This case seems to have been just an accident, but fears of sabotage were common to ship and weapons construction during

wartime. Some of these fears appeared to be justified in at least one case, the incomplete and ill-starred CSS *Mississippi* at New Orleans, when workers apparently found a wooden pin driven into the launching ways. This was believed to have prevented the *Mississippi*'s successful launch on the first attempt, before the pin was discovered. As a strategic and cosmopolitan port city, New Orleans was probably especially rife with spies and potential saboteurs. The incident during the *Mississippi*'s construction is related in detail in Kean, *Report of Evidence*, 114.

112. Robert F. Pinkney, report to William W. Hunter, August 3, 1863, folder 12, box 1, Savannah Squadron Papers.

113. Driver, *Confederate Sailors*, 342–43. Nothing appears to be known about Tynan outside of his Confederate service. He seems to have finished the war on the cruiser CSS *Tallahassee*, but there are no records of him afterward.

114. John M. Tynan, report to William W. Hunter, August 4, 1863, folder 12, box 1, Savannah Squadron Papers.

115. Maurice Melton, *The Best Station of Them All: The Savannah Squadron, 1861–1865* (Tuscaloosa: University of Alabama Press, 2012), 245. Two weeks was an amazingly short turnaround time, especially considering the wartime exigencies of shipping the disassembled machinery to the other side of Georgia, completely recasting several major components, and then shipping it back.

116. Driver, *Confederate Sailors*, 268. Pinkney later commanded the naval defenses of North Carolina until the war's end, when he was paroled at Greensboro, North Carolina. He lived in Baltimore until his death in March 1878.

117. R. Thomas Campbell, *Southern Service on Land and Sea: The Wartime Journal of Robert Watson, CSA/CSN* (Knoxville: University of Tennessee Press, 2002), 111.

118. Ibid., 139–40. The "shell" may have been a solid iron bolt, used by both sides to smash armor plate.

119. Judy Wood, e-mail to author, October 26, 2011.

Chapter 5. The Tennessee Class Ironclads

1. Holcombe, "Evolution," 68–69.
2. Still, *Confederate Navy*, 67.
3. Ibid., 66–67; Still, *Confederate Shipbuilding*, 38.
4. Still, *Iron Afloat*, 192, 195.
5. Holcombe, "Evolution," 71–72; US Navy, *Official Records*, ser. 1, 21:547.
6. Way, *Packet Directory*, 16.
7. US Navy, *Official Records*, ser. 1, 21:547–50.
8. Way, *Packet Directory*, 468.
9. US Navy, *Official Records*, ser. 1, 21:548–49.
10. Way, *Packet Directory*, 468; Still, *Iron Afloat*, 191–92.
11. US Navy, *Official Records*, ser. 1, 20:735.
12. Ibid., ser. 1, 21:548–49.
13. Ibid. This is the only instance of multiple boilers with a common furnace

on a ship that I have come across in the records. The usual situation is multiple furnaces per boiler.

14. Still, *Confederate Navy*, 76–79.

15. Ralph W. Donnelly, "The Charlotte, North Carolina, Navy Yard, CSN," *Civil War History* 5 (1959): 73.

16. Ibid., 176–177. Immediately before the war's end, Jones was in command of CSS *Stockdale* at Mobile. He was paroled there in May 1865 and spent his last years as a farmer in Alabama. He was shot and killed on June 21, 1877.

17. Violet Alexander, "The Confederate States Navy Yard at Charlotte, N.C., 1862–1865," *Southern Historical Society Papers* 40, no. 2 (1915): 185.

18. Donnelly, "Charlotte [NC] Navy Yard," 74.

19. Alexander, "Confederate States Navy Yard," 186; Donnelly, "Charlotte [NC] Navy Yard," 77. Alexander states that a torpedo (mine) shop and testing laboratory were also located at the site.

20. Donnelly, "Charlotte [NC] Navy Yard," 77.

21. Alexander, "Confederate States Navy Yard," 192; Donnelly, "Charlotte [NC] Navy Yard," 77. According to Alexander, Dwyer's machine was confiscated by the US government and put into general use in US navy yards after the war. Dwyer was apparently never compensated financially (perhaps this is understandable, since he was a rebel) or given any credit for such an important invention.

22. Alexander, "Confederate States Navy Yard," 186; Donnelly, "Charlotte [NC] Navy Yard," 78. Alexander also lists "the gun-boats built at Charleston and Savannah," the never completed ironclad *Mobile* (almost nothing is known of its intended configuration), and "other ironclads built at New Orleans." This last statement appears to be in error, since New Orleans had fallen well before the Charlotte works became operational. Chief Engineer Ramsay himself wrote that ironclads at New Bern (North Carolina), Wilmington, Charleston, Savannah, Columbus, Mobile, and New Orleans had shafting forged at Charlotte. Ramsay, "Narrative," frame 0352.

23. Donnelly, "Charlotte [NC] Navy Yard," 79. Charlotte could also be called the last Confederate capital, since Jefferson Davis and his cabinet stayed there a few days after evacuating Richmond.

24. Driver, *Confederate Sailors*, 175. Johnston (paroled after his capture on the *Tennessee*) was with the last Confederate naval forces in Alabama when they surrendered at Nanna Hubba Bluff on the Tombigbee River in May 1865. He died in Savannah, Georgia, on May 9, 1896, at age seventy-nine.

25. Ibid., 44–45. After the war Buchanan was president of the Maryland Agricultural School in 1868–1869. He died in May 1874.

26. Johnson and Buell, *Battles and Leaders*, 4:405.

27. Melton, *Confederate Ironclads*, 214–15.

28. Ibid., 216.

29. J. D. Johnston, "Admiral Buchanan and the Confederate States Ram 'Tennessee,'" *United Service* 7, no. 2 (1882): 208.

30. Melton, *Confederate Ironclads*, 218–20.

31. Ibid., 221–23. Perhaps the *Tennessee*'s greatest weakness was its steering system, in which the steering chains themselves were left vulnerable. They ran from the casemate in shallow trenches across the aft flush deck, covered only with boiler iron. Commander Johnston was well aware of this, but nothing had been done about it by the time of battle.

32. Mooney, *Dictionary*, 2:574.

33. Holcombe, "Evolution," 74.

34. Ibid., 72, 88.

35. Still, *Confederate Shipbuilding*, 60.

36. Holcombe, "Evolution," 74.

37. Ibid., US Navy, *Official Records*, ser. 2, 1:251.

38. Holcombe, "Evolution," 73.

39. Turner, *Navy Gray*, 158.

40. US Navy, *Official Records*, ser. 2, 1:251.

41. Turner, *Navy Gray*, 223.

42. Holcombe, "Evolution," 88–89.

43. Flanders and Westfall, *Memoirs of E. A. Jack*, 35.

44. Driver, *Confederate Sailors*, 115. Freeman was a civil engineer in his native Norfolk after the war.

45. Flanders and Westfall, *Memoirs of E. A. Jack*, 35.

46. Ibid. This account is particularly valuable because it explains why the *Columbia* was wrecked. Other sources simply state that the ironclad hit a sunken vessel.

47. Ibid.

48. Ibid., 35–36. Jack afterwards wound up on CSS *Richmond* of the James River Squadron. After his ship was destroyed and the Confederate capital evacuated, he was captured on April 6, 1865, and imprisoned in Washington, DC. Jack later served briefly with the revolutionary forces of Colombia when that nation became independent of Spain, but eventually he returned to the Norfolk area. He also had a brief career in the US Revenue Cutter Service. Jack died on December 18, 1911. Driver, *Confederate Sailors*, 168–69.

49. US Navy, *Official Records*, ser. 1, 15:373; Mooney, *Dictionary*, 2:510.

Chapter 6. Other Standard Hulls

1. Driver, *Confederate Sailors*, 130. Graves completed the *Charleston* and the *Virginia II*, but ironclads of his design were also laid down at Oven Bluff, Alabama, and at Savannah, Georgia. In addition, he began construction on a double-ender ironclad (propulsion machinery at both ends and a short casemate with one gun amidships) at Richmond. All these were destroyed on the stocks before completion. Holcombe, "Evolution," 81–82, 131.

2. Wexler, "Ordinary Gunboats," 64.

3. Holcombe, "Evolution," 78.

4. Ibid.

5. Ibid., 90. The models were supposedly taken from Richmond in April 1865, possibly from the navy offices themselves.

6. Ibid., 77–79.

7. Melton, *Confederate Ironclads*, 165–66.

8. Holcombe, "Evolution," 81. The 8-inch thickness is stated by enough primary sources that it must be considered accurate, especially considering the previous wartime experience of ironclad combat.

9. Coski, *Capital Navy*, 86, 237.

10. Holcombe, "Evolution," 90.

11. Coski, *Capital Navy*, 211.

12. US Navy, *Official Records*, ser. 2, 1:269; Dew, *Ironmaker*, 122, 265; Coski, *Capital Navy*, 72.

13. Still, *Confederate Navy*, 67.

14. William Radford, report to US Navy Department, April 25, 1865, PN file, roll 42, ONRL, RG 45. Radford's report listed in detail many parts and sets of machinery, including that intended for the unfinished *Tennessee* class ironclad the *Texas*. Its engines were British-built and probably taken from a wrecked blockade-runner of the *Don* class.

15. *Virginia II* plans, 82–12–2E, Bureau of Ships, Records, 1794–1972, Record Group 19, National Archives and Records Administration, College Park, MD.

16. US Navy, *Official Records*, ser. 1, 9:801.

17. Coski, *Capital Navy*, 254.

18. US Navy, *Official Records*, ser. 1, 10:186. There is again the possibility of sabotage, but no one seems to have considered it at the time. At least, there is no mention of it in Mitchell's report.

19. Ibid., 718–19.

20. Coski, *Capital Navy*, 207.

21. US Navy, *Official Records*, ser. 1, 12:179.

22. Coski, *Capital Navy*, 180.

23. Nelson D. Lankford, ed., *An Irishman in Dixie: Thomas Conolly's Diary of the Fall of the Confederacy* (Columbia: University of South Carolina Press, 1988), 65–66.

24. Holcombe, "Evolution," 83. One of these vessels, built at Selma, Alabama, was apparently severely damaged at launch, purchased for conversion to a blockade-runner named the *Phoenix*, and later scuttled as an obstruction in Mobile Bay. The wreck site has been studied, and the vessel appears to be a standard hull. It is currently uncertain how closely the *Phoenix* may resemble the *Nashville*. David A. Ball, "*Phoenix*, and the Confederate Obstructions of Upper Mobile Bay, Alabama" (master's thesis, Florida State University, Tallahassee, 1998), 87.

25. Holcombe, "Evolution," 83, 91.

26. Still, *Iron Afloat*, 195.

27. Holcombe, "Evolution," 83. *Hog* is the term used to describe the way the bow and stern of a vessel can droop because of structural weakness in the midsection or from the ends being too heavy.

28. Ibid., 82–83.

29. Janet B. Hewett, ed., *Supplement to the Official Records of the Union and*

Confederate Armies (Wilmington, NC: Broadfoot, 1999), 3:320; US Navy, *Official Records*, ser. 1, 20:735.

30. US Navy, *Official Records*, ser. 1, 25:653.

31. Way, *Packet Directory*, 302.

32. Way, *Packet Directory*, 302. Way lists the length as 265 feet and the draft as 8.5 feet. These should probably be considered simple transcription errors.

33. Way, *Packet Directory*, 71.

34. James H. Warner, order to William A. Campbell, July 25, 1864, James Warner Letterbook, Port Columbus National Civil War Naval Museum, Columbus, GA; Voucher to John Lawson, September 30, 1863, AC file, roll 6, ONRL, RG45; D. DeHaven, contract with R. B. Powell, April 20, 1863, PN file, roll 42, ONRL, RG45.

35. US Navy, *Official Records*, ser. 1, 22:225; Hewett, *Supplement to the Official Records*, 3:320.

36. Hewett, *Supplement to the Official Records*, 3:320.

37. Melton, *Confederate Ironclads*, 206, 214, 224–25. It is doubtful that the *Nashville* could have taken the full weight of armor even if enough had been available for total covering. It was also fortunate that the ironclad never encountered the stresses of open water; Union Admiral Thatcher, for one, was certain that the *Nashville* could not live in a seaway. The Confederates were probably lucky that the ship was not tested against enemy warships as CSS *Tennessee* was.

38. Ibid., 224–25.

39. Driver, *Confederate Sailors*, 25; Melton, *Confederate Ironclads*, 225.

40. Melton, *Confederate Ironclads*, 225–27.

41. Ibid., 230–32.

42. Mooney, *Dictionary*, 2:552.

Chapter 7. Early Attempts at an Alternative Hull Form

1. Holcombe, "Evolution," 93, 95.

2. Still, *Iron Afloat*, 192.

3. Holcombe, "Evolution," 95–96.

4. Ibid., 93.

5. Still, *Iron Afloat*, 191.

6. Turner, *Navy Gray*, 145.

7. Continental Eagle Corporation, "Factory at War," Pratt History, http://www.pratthistory.com/factory_at_war.htm. This webpage has been inaccessible since May 2014. Continental Eagle Corporation of Prattville occupies the old site of William Penny & Company and may have access to pertinent records.

8. Bert Neville, *Directory of Steamboats with Illustrations and Lists of Landings on the Chattahoochee-Apalachicola-Flint-Chipola Rivers* (Selma, AL: Coffee, 1961), 38.

9. Turner, *Navy Gray*, 162.

10. Holcombe, "Evolution," 94.

11. Franklin Buchanan to Stephen R. Mallory, April 6, 1863, 156, Franklin

Buchanan Letterbook, Buchanan and Screven Family Papers, 1801–1913, Southern Historical Collection, University of North Carolina, Chapel Hill.

12. Still, *Iron Afloat,* 224–25. This is one of very few instances in which the Confederates simply sank the vessel instead of blowing it up, which accounts for the intact state of the wreck. The *Huntsville* was dispatched in a similar manner nearby.

13. Holcombe, "Evolution," 95. Having been present at the excavations, Holcombe attests to the extremely poor visibility and large amounts of river mud covering the wrecks of the *Tuscaloosa* and the *Huntsville.* Only a small portion of the casemate of one and the bow of the other were uncovered during the project.

14. Still, *Iron Afloat,* 191.

15. US Navy, *Official Records,* ser. 1, 21:363.

16. Way, *Packet Directory,* 250.

17. Ibid.; Edward A. Mueller, *Perilous Journeys: A History of Steamboating on the Chattahoochee, Apalachicola, and Flint Rivers, 1828–1928* (Eufaula, AL: Historic Chattahoochee Commission, 1990), 102.

18. Larry J. Daniel and Riley W. Gunter, *Confederate Cannon Foundries* (Union City, TN: Pioneer Press, 1977), 34; Warren Ripley, *Artillery and Ammunition of the Civil War,* 4th ed. (Charleston, SC: Battery Press, 1984), 181.

19. Franklin Buchanan to Stephen R. Mallory, May 13, 1863, 171, Buchanan Letterbook.

20. Still, *Iron Afloat,* 224. Still notes that there is no mention of the *Tuscaloosa* participating in the defense of Spanish Fort; therefore, the ironclad might not have been operational at that time. Mechanical difficulty was probably, as always, the culprit.

21. Allen R. Saltus and Sidney H. Schell, *The CSS Huntsville and CSS Tuscaloosa Project, Mobile River, Alabama: Report of 1985 Activities* (Mobile, AL: US Army Corps of Engineers and Mobile Jaycees, 1985), 16–17. After the fall of Mobile, the remaining Confederate forces retreated upriver. The water was high from heavy spring rains, and the *Tuscaloosa* and the *Huntsville* could not stem the current, so they were abandoned. According to Saltus and Schell, "The *Huntsville* is reported to have been scuttled by cutting the main feedwater line into her engine room, allowing the vessel to fill and sink. The *Tuscaloosa* was probably scuttled in a manner similar to the *Huntsville,* the vessels sinking in about 24 feet of water."

22. Still, *Iron Afloat,* 101.

Chapter 8. Diamond Hull Ironclads

1. Holcombe, "Evolution," 99–100. Holcombe points out that despite the miscalculations, the drafts of these ironclads were still much shallower than other designs.

2. Ibid., 96–99.

3. Ibid., 96.

4. Robert G. Elliott, *Ironclad of the Roanoke: Gilbert Elliott's Albemarle* (Shippensburg, PA: White Mane, 2005), 11–12, 17, 40–41, 55–57, 62–63. As far as is

known, all the vessels Elliott contracted to build for the Confederate government were designed by Constructor Porter. The gunboats at Elizabeth City appear to have been of the same basic design as CSS *Chattahoochee*, and the Tarboro ironclad was probably a sister to the *Albemarle*. Very little is known about the Dismal Swamp ironclads at Norfolk.

5. Ibid., 64–65, 104, 139–40. The command of the *Albemarle* was the highlight of Cooke's wartime career. After the famous battle, he resigned the command for health reasons. He later requested command of the naval facilities at Halifax and Edward's Ferry, and after an initial denial (the post had already been filled), he received that appointment when the previous commander was reassigned. Cooke finished the war at Halifax, apparently in bad health most of the time.

6. Ibid., 115, 164, 166.

7. Holcombe, "Evolution," 99.

8. Elliott, *Ironclad of the Roanoke*, 141–42. Loyall and several other ex-Confederates were at a reunion in Norfolk on January 19, 1897, celebrating the late Robert E. Lee's birthday.

9. H. W. Wilson, *Ironclads in Action: A Sketch of Naval Warfare from 1855 to 1895, with Some Account of the Development of the Battleship in England*, 4th ed. (London: Sampson Low, Marston, 1896), 1:107, https://play.google.com/store/books/details?id=RHDVAAAAMAAJ.

10. Scharf, *History*, 404; Elliott, *Ironclad of the Roanoke*, 404.

11. John N. Maffitt, "Reminiscences of the Confederate Navy," *United Service* 3, no. 1 (1880): 503.

12. Elliott, *Ironclad of the Roanoke*, 143.

13. Ibid., 159–67.

14. Ibid., 167–68.

15. Ibid., 178–83. Only nine iron plates were cracked. The person killed, a crew member by the name of Harris, had briefly stuck his head out of one of the gunports to glimpse the battle with USS *Miami*. A sailor on that vessel dispatched Harris with a quick pistol shot.

16. Mooney, *Dictionary* (1959), 1:495.

17. John Stephen Carbone, *The Civil War in Coastal North Carolina* (Raleigh: North Carolina Department of Cultural Resources, Division of Archives and History, 2001), 116–17.

18. Ibid.

19. Elliott, *Ironclad of the Roanoke*, 228–229. Maffitt was born in 1819 and died in 1886 in Wilmington, North Carolina. After commanding the *Albemarle*, he commanded the blockade-runner *Owl*. Driver, *Confederate Sailors*, 216.

20. Elliott, *Ironclad of the Roanoke*, 245.

21. Stickney, *Promotion or the Bottom*, 149–50.

22. Elliott, *Ironclad of the Roanoke*, 249–51.

23. Stickney, *Promotion or the Bottom*, 151–53. It is amazing that none of Cushing's men were killed by the small-arms fire or grapeshot fired from the *Albemarle*'s forward 6.4-inch Brooke rifle. Only two men died by drowning when the launch was wrecked; eleven others were captured.

24. Ibid., 153.

25. Ibid., 154. Warley reported the sad news to Captain Cooke at Halifax, then moved on to Wilmington. He did not stay long, for there was no ship there to command, but he returned to Charleston in time for the destruction of the Charleston Squadron. Warley finished the war in the army and was captured at Athens, Georgia, in May 1865. Afterward he settled in the city of his greatest exploits, New Orleans, and lived out his days as a clerk at the Board of Liquidation, City Debt. Alex Warley died on January 12, 1895, and was buried in his native South Carolina.

26. Elliott, *Ironclad of the Roanoke*, 273.

27. Ibid., 273–78. Several artifacts remain today in various places, including the shell-riddled smokestack, one of the 6.4-inch Brooke rifles, and the ship's bell.

28. Ibid., 267, 274.

29. Holcombe, "Evolution," 100.

30. US Department of War, *The War of the Rebellion: A Compilation of the Official Records of the Union and Confederate Armies* (1880–1901; repr., Harrisburg, PA: National Historical Society, 1985), ser. 1, 33:290; Leslie S. Bright, William H. Rowland, and James C. Bardon, *CSS Neuse: A Question of Iron and Time* (Raleigh: North Carolina Department of Cultural Resources, 1981), 8, 150.

31. Bright et al., *CSS Neuse*, 8. An attempt was made by a Union soldier to set fire to the uncompleted hull, but he was unsuccessful. Most damage to the ship came from shells fired across the river during the brief but intense fighting around Whitehall on December 16. The raiders also destroyed a large quantity of supplies and construction materials.

32. Ibid., 12.

33. Still, *Confederate Navy*, 67–68.

34. Lawrence W. Sagle and Alvin F. Staufer, *B&O Power: Steam, Diesel, and Electric Power of the Baltimore and Ohio Railroad, 1829–1964* (N.p.: A. F. Staufer, 1964), 36, 41. Jackson's capture of that vital railway junction and yard was a windfall for the machinery-starved Confederacy: eleven locomotives and the cars with them. In a feat of strength and ingenuity, five engines (including no. 34) were hauled overland south into Virginia, and the others were stripped of everything usable.

35. US War Department, *War of the Rebellion*, ser. 1, 33:290. The sheer volume of information in both this collection and that of the Navy official records is still capable of revealing unknown details like this to the modern researcher. The machinery installation at Pugh's Mill was probably typical of small rural sawmill and factory operations throughout the United States in the mid-nineteenth century.

36. Driver, *Confederate Sailors*, 202–3. Benjamin Pollard Loyall (1833–1923) hailed from Norfolk and graduated first in his class from the US Naval Academy in 1855. He was a lieutenant when he resigned his commission in 1861 and was appointed first lieutenant in the Confederate navy. Loyall's Civil War career included service at the Gosport Naval Yard, at Roanoke Island, and on CSS *Richmond* and CSS *Patrick Henry* on the James River before his duty in North Carolina and as captain of the *Neuse*. Loyall finished the war with gallantry and afterward lived prosperously in his native Norfolk.

37. Still, *Iron Afloat*, 162.

38. Bright et al., *CSS Neuse*, 15.

39. Still, 221.

40. Ibid.

41. William R. Trotter, *Ironclads and Columbiads: The Civil War in North Carolina, the Coast* (Winston Salem, NC: John F. Blair, 1989), 410. The actual scuttling took place on March 12, 1865.

42. Peter B. Campbell, "The Development of Confederate Ship Construction: An Archaeological and Historical Investigation of Confederate Ironclads *Neuse* and *Jackson*" (master's thesis, East Carolina University, Greenville, NC, 2009), 58.

43. W. L. Poalk, "Advertisement," *North Carolina Daily Times*, October 3 1865; Campbell, "Development," 58.

44. Bright et al., *CSS Neuse*, 153.

45. Holcombe, "Evolution," 102–3. Cooke wrote that the *Albemarle* could have achieved a complete victory with two more guns, one in each broadside. An improved design incorporating Cooke's suggestions and perhaps being of a configuration similar to that of the *Fredericksburg* was actually laid down at Edward's Ferry by Gilbert Elliott in August 1864 but was never completed.

46. Ibid., 103.

47. Coski, *Capital Navy*, 81.

48. Holcombe, "Evolution," 103–5.

49. Dew, *Ironmaker*, 265; Still, *Confederate Navy*, 68; Coski, *Capital Navy*, 72.

50. Coski, *Capital Navy*, 85–86.

51. Ibid., 200–202. The *Fredericksburg*'s captain during the Trent's Reach battle, Lieutenant Francis E. Shepperd, was an aggressive young officer who did everything he could to harass the enemy and garner glory for himself and his ship. Like many a commander before him, he was hampered by overcautiousness on the part of his superiors and the limitations of his command. A run through the gauntlet of obstructions, shore batteries, and enemy warships (including monitors) such as CSS *Arkansas* had done would probably have destroyed the *Fredericksburg*. Ironclads were not quite as invincible in 1865 as they were in 1862.

52. Holcombe, "Evolution," 105.

53. Katherine Brash Jeter, "Against All Odds: Lt. Jonathan H. Carter, CSN, and His Ironclad," *Louisiana History* 28, no. 3 (1987): 267–68.

54. Holcombe, "Evolution," 108. The Trans-Mississippi Department especially suffered from a lack of these essential shipbuilding requirements.

55. Still, *Iron Afloat*, 101–2; Holcombe, "Evolution," 105–9.

56. Holcombe, "Evolution," 108.

57. Ibid. Many photographs of these vessels in service are found in US Department of the Navy, Photo Archive, Naval History and Heritage Command, http://www.history.navy.mil/our-collections/photography.html.

58. Robert M. Sheppard III, "U.S.S. *Cairo* Engines and Boilers," American Society of Mechanical Engineers, 1990, 3, https://www.asme.org/getmedia/2306f17b-2b7f-4592-bea8-073cf64700fa/143-USS-Cairo-Engine-and-Boilers-1862.aspx.

59. Holcombe, "Evolution," 107–8.

60. Way, *Packet Directory*, 196, 386.

61. A. Robert Holcombe Jr., e-mail to author, April 17, 2012.

62. Way, *Packet Directory*, 443.

63. "The T. W. Roberts," *Louisville Daily Courier*, November 19, 1860, 4; Way, *Packet Directory*, 443. Way lists the boat as being 158 feet long with a draft of 6.1 feet. The differences in measurements between the *Missouri*'s machinery and the *T. W. Roberts*'s (if in fact that boat was the source) could perhaps be accounted for by errors in measurement, or inside versus outside cylinder diameter, a common distinction often neglected in engine measurements. The difference of 1 foot in stroke length and four versus three boilers are more difficult to account for.

64. Ibid., 196, 443; Paul H. Silverstone, *Warships of the Civil War Navies* (Annapolis, MD: US Naval Institute Press, 1989), 247.

65. Way, *Packet Directory*, 196.

66. Katherine Brash Jeter, *A Man and His Boat: The Civil War Career and Correspondence of Lieutenant Jonathon H. Carter, CSN* (Lafayette: Center for Louisiana Studies, 1996), 6, 37.

67. Ibid., 6; A. Robert Holcombe Jr., e-mail to author, April 25, 2012.

68. Still, *Confederate Navy*, 105.

69. Jeter, "Against All Odds," 263–88.

70. Driver, *Confederate Sailors*, 55–56. After the war Carter seems to have lived quietly in Virginia, where he died in the town of Shirley in 1887.

71. Jeter, *Man and His Boat*, 6, 31.

72. Ibid., 39–41, 49–50. He was correct, although the *Missouri* was faster than expected. All the Union City class ironclads and other centerwheel vessels such as USS *Benton* and USS *Essex* were noted by contemporaries as slow, even for ironclads. Since all the centerwheelers on both sides served on the Mississippi River system, fighting swift currents was part of the problem.

73. Ibid., 51, 54–55. This is the most detailed account of a Confederate captain dealing with the machinery installation on his vessel that I have found. Usually the deck officers, perhaps to no surprise, have little to say on engineering matters in their memoirs and contemporary accounts.

74. Ibid., 67, 68, 70–71.

75. Ibid., 78–79.

76. Ibid., 150. Tenders were commonly used, since many steamers (especially heavy ironclads) burned fuel at a prodigious rate and did not have enough room for a supply of more than a few days. At least one other Confederate ironclad, CSS *Albemarle*, is known to have had a tender in attendance during operations.

77. Ibid., 175.

78. Ibid., 183–84.

79. Still, *Iron Afloat*, 226.

80. Jeter, "Against All Odds," 287.

81. Sheppard, "U.S.S. *Cairo*," 6.

82. Jeter, "Against All Odds," 287–88.

Chapter 9. Uncompleted Vessels

1. Holcombe, "Evolution," 1.

2. Still, *Iron Afloat*, 43.

3. Holcombe, "Evolution," 44, 58.

4. Kean, *Report of Evidence*, 125. Design simplicity was such that ordinary house carpenters were used in the construction of the vessel.

5. Holcombe, "Evolution," 44, 46. Sinclair's testimony is in US Navy, *Official Records*, ser. 2, 1:489; Farragut's is in ibid., ser. 1, 18:158.

6. Robert Minor, journal entry, January 30, 1862, Robert Minor Journal, Minor Family Papers, 1810–1932, Virginia Historical Society, Richmond. Minor was one of the Confederate navy's most gifted technical experts. He was born on September 12, 1827, in Virginia, was a Mexican War veteran, and ranked as a first lieutenant when he resigned his commission in April 1861. He was quickly given the same rank in Confederate service and commanded CSS *St. Nicholas* on the Potomac River until his appointment as the flag lieutenant to Franklin Buchanan on CSS *Virginia*. Minor afterward commanded the Naval Ordnance Works in Richmond and finished the war as a flag lieutenant and an ordnance officer of the James River Squadron. After the war he had a varied and eventful career building vessels for the Chilean navy, as a businessman in Richmond, and as an engineer for the James River Improvement Company. Minor died on November 25, 1871. Driver, *Confederate Sailors*, 233–34.

7. Still, *Iron Afloat*, 44.

8. Holcombe, "Evolution," 44.

9. Kean, *Report of Evidence*, 155, 159.

10. John Roy, diary entry, September 20, 1862, John Roy Diary, 1860–1862, Howard-Tilton Memorial Library, Tulane University, New Orleans, LA.

11. US Navy, *Official Records*, ser. 2, 1:582–83; Holcombe, "Evolution," 45.

12. Holcombe, "Evolution," 45.

13. Kean, *Report of Evidence*, 155–56. In yet another "could have been" regarding the *Mississippi*, a speed of 14 knots would have ranked it among the fastest ships afloat.

14. Merrill, "Confederate Shipbuilding," 90–91.

15. Luraghi, *History*, 123.

16. Kean, *Report of Evidence*, 128.

17. Ibid., 112.

18. US Navy, *Official Records*, ser. 2, 1:577–78.

19. Ibid. One wonders if similar weak castings, left unnoticed and worked too hard, were a cause of CSS *Arkansas*'s career-ending engine problems.

20. Kean, *Report of Evidence*, 381.

21. US Navy, *Official Records*, ser. 2, 1:572.

22. Merrill, "Confederate Shipbuilding," 90. Unfortunately, little information has been found on the Patterson and Clark works. The histories and specialties of those establishments may be set forth in some local history work in New Orleans.

23. US Navy, *Official Records*, ser. 2, 1:535.

24. Still, *Iron Afloat*, 59. There are some contemporary sketches of the *Mississippi* burning in front of New Orleans by Union war artist and correspondent William Waud.

25. US Navy, *Official Records*, ser. 2, 1:596. Whether the ship really would have been operational at that time is another matter. Virtually every Confederate ironclad actually completed took several months to work up after its launch.

26. Holcombe, "Evolution," 46.

27. Merrill, "Confederate Shipbuilding," 93; Still, *Confederate Navy*, 30–31.

28. Holcombe, "Evolution," 109–10.

29. Turner, *Navy Gray*, 170, 176, 214.

30. Holcombe, "Evolution," 110–11.

31. Augustus McLaughlin, log entry, July 13, 1864, Navy Yard Log, Augustus McLaughlin Letterbook, 1864–1865, Port Columbus National Civil War Naval Museum, Columbus, GA.

32. Neville, *Directory of Steamboats*, 42.

33. George F. Pearce, *Pensacola during the Civil War: A Thorn in the Side of the Confederacy* (Gainesville: University Press of Florida, 2000), 117.

34. Braxton Bragg, order to George G. Garner, October 21, 1861, file T-5, roll 29, Microfilm Publication M909, US Department of War, Confederate Records, 1825–1900, Record Group 109, National Archives and Records Administration, College Park, MD.

35. Turner, *Navy Gray*, 176.

36. Campbell, "Development," 82.

37. Ibid., 83.

38. Turner, *Navy Gray*, 237. The *Jackson*'s destruction by enemy forces ran counter to the usual wartime fate of Confederate ironclads being blown up by their own crews.

39. Ibid., 235–37.

40. Campbell, "Development," 79.

41. Melton, *Best Station*, 65–66. Willink was a shipbuilder's son with some formal schooling but left Georgia to learn his father's trade in New York. He worked for nine years at the famous William H. Webb shipyard before returning to take over his father's yard at Savannah. During the Civil War he made a substantial contribution to Confederate ship construction.

42. Holcombe, "Evolution," 123, 125; *Milledgeville* Plan File CX1, Savannah Squadron Papers.

43. Holcombe, "Evolution," 126.

44. Ibid., 123.

45. Mooney, *Dictionary*, 2:549.

46. Melton, *Best Station*, 388.

47. Judy Wood, e-mail to author, October 26, 2011.

48. Holcombe, "Evolution," 127, 130.

49. Ibid., 127–130.

50. J. H. Warner, report to J. H. Alexander, June 17, 1865, case file 825, "Captured and Abandoned Property," US Department of the Treasury, General

Records, 1789–1990, Record Group 56, National Archives and Records Administration, College Park, MD.

51. Holcombe, "Evolution," 130.

52. Warner report to Alexander, June 17, 1865. Warner also lists several pairs of engines, boilers, propellers, and components intended for other unfinished ironclads, including one at Charleston and one of the Bigbee boats on the Tombigbee River in Alabama. He also lists one pair of large engines manufactured by "Messrs. Fawcett & Preston of Liverpool . . . imported through the blockade." The question of which vessels used or were intended to use British-built machinery thus arises again.

Chapter 10. Results and Overview

1. Still, *Iron Afloat*, 227.

2. C. H. McBlair, "A Historical Sketch of the Confederate Navy," *United Service* 3, no. 1 (1880): 594.

3. Luraghi, *History*, 346.

4. Still, *Iron Afloat*, 228.

5. Ibid., 102–4.

6. Luraghi, *History*, 10.

7. Quoted in Still, *Iron Afloat*, 227.

GLOSSARY

The coherence of this glossary is entirely a result of the author's extensive consultation of Alan Bates, *The Western Rivers Engineroom Cyclopoedium* (Louisville, KY: Cyclopoedium Press, 1996).

Ironclad Terms

Ironclads were a very new type of warship in the 1860s that required their own set of anatomic terminology, which in many cases were developed from terms describing wooden vessel construction. They are reproduced here, adapted from the groundbreaking work of A. Robert Holcombe Jr., "The Evolution of Confederate Ironclad Design" (master's thesis, East Carolina University, Greenville, NC, 1993), 149–50.

Berth deck: A deck below the gun deck and the flush decks that contained the crew's quarters. Separated by the engine and boiler rooms, the forward berth deck was usually reserved for enlisted personnel, and the aft berth deck was for the wardroom and officers' quarters.

Casemate (citadel, shield): An armor-plated enclosure, commonly sloped on the sides and ends, containing guns that fired through ports.

Depth of hold: On virtually all purpose-built Confederate ironclads, a vertical measurement made from the top of the gun deck beam at the side of the hull to the lower edge of the keel rabbet, taken at the midship section. On an ironclad with no keel, the bottom of the floor planks was used as the lower point of reference.

Flush deck: A low, exposed deck located both fore and aft of the casemate, usually an extension of the gun deck within the casemate.

Gun deck: A deck within the casemate on which guns were mounted, often on the same level as the exposed fore and aft flush decks.

Knuckle (eaves, sponson): (1) An angle formed by the junction of the casemate sides and built on the protrusion from the hull of standard hull and similar ironclads; (2) an angle formed by the junction of the casemate sides and the flaring hull of a diamond hull ironclad.

Length between perpendiculars: On a standard hull ironclad and most other designs, the measurement taken between the aft side of the stem at the level of the flush deck (fore perpendicular) and the fore side of the stern post at the level of the flush deck (aft perpendicular). On a diamond hull ironclad, because of the usual raking stem, this measurement was most often taken from the forward edge of the fore flush deck (usually underneath the armor) to the fore side of the stern post at the level of the flush deck. A notable exception was the centerwheel ironclad plan, where the measurement was taken between the intersection of the knuckle and aft side of the stem and the extreme aft end of the hull.

Molded beam: On a standard hull ironclad and designs with built-on knuckles, the width of the hull at its widest point measured across the futtocks but not including the hull planking or knuckles. On a diamond hull ironclad and designs without built-on knuckles, the measurement was made across the junction of the hull and casemate frames, usually not including the planking or armor plate.

Pilot hatch: Positioned on the spar deck, an open hatch with 9-inch-high sloped combings within which the pilot, mounted on a ladder with the ship's wheel on the gun deck below, peered from and conned the vessel. It was common to several early ironclads, including the *Savannah*.

Pilothouse: A small truncated pyramid about 3 feet high on the spar deck, equipped with viewing slits or ports through which the pilot standing inside guided the ship. Although the pilothouse was most often used singularly, some ironclads, like the *Fredericksburg*, had one mounted on each end of the spar deck.

Ram: A protrusion on the forward end of a vessel used for piercing the side of an enemy ship. On early ironclads the ram was usually formed by a heavy, pointed iron piece bolted to the stem. On many standard hull ironclads, particularly those designed by John L. Porter, the ram was formed by a forward extension of the knuckle; thus it was an integral part of the ship's structure and was not likely to be broken off. A ram on a diamond hull ironclad was formed by a wooden extension of the stem covered with one or more layers of iron plate.

Spar deck (casemate deck, shield deck): The exposed weather deck forming the top of the casemate, usually equipped with hatches to the gun deck below, and through which the smoke pipe and other appurtenances passed or were attached.

Steam Machinery Terms

Steam technology, once ubiquitous throughout the world, is fast becoming as obscure in its daily operation and terminology as the horse and sail-powered transportation systems it initially competed with. Therefore this glossary attempts to list the terms for all major machinery components presented in the text. A perusal here, it is hoped, will lead to a more complete understanding and a greater appreciation of the technology itself and the seemingly superhuman engineers running it in such adverse conditions.

BOILERS

Ash chute: A tube used for the removal of boiler furnace waste products. On a riverboat it commonly goes through the bottom of the hull forward of the furnaces; an oceangoing vessel has a more elaborate system to haul ashes up one or more decks for discharge over the side.

Boiler: The centerpiece of any external combustion (steam) system. It consists of one or more furnaces burning coal or wood fuel; the heat produced is transmitted by exhaust gases to the surrounding water encased in the shell, producing steam. Three main types were used on Confederate ironclads: flue, firetube, and watertube.

Breeching: A connection between the boiler flue or tube and the smokestack.

Draft (draught): Airflow created by the movement of hot gases upward in a confined space. Constant removal of these gases draws in more air below, aiding in combustion. Confederate ironclads used two kinds of draft: natural and forced. The former is what it sounds like; see *fan blower* below (under Auxiliary Equipment) for the latter. This is not to be confused with the depth at which at a vessel sits in the water.

Fire front: The end of the boiler where the furnaces are located.

Firetube: A boiler type in which exhaust gases pass through many small tubes to heat the water reservoir around them.

Flue: A boiler type in which exhaust gases pass through two or more large manhole–size tubes, or flues, to heat the water reservoir around them. It is used almost exclusively on riverboats.

Furnace: The area of the boiler where combustion occurs. Firemen load fuel, tend fires, and perform most of their work here to provide adequate heat for maintaining steam.

Grate (grating): Metal bars or plates arranged in the furnace to hold fuel and allow appropriate air circulation for ideal combustion. Different grates are used for wood and coal.

Heating surface: The total heated area in the boiler to which water is exposed for steam production.

Main steam line: A large pipe leading from the steam drum for supplying steam to the engines.

Mud drum: A hollow metal cylinder, usually mounted transversely beneath the after end of a boiler set, used for collecting any mud or other debris that may be present in boiler water. It is most commonly encountered on noncondensing vessels such as riverboats.

Mud valve: An apparatus located on the mud drum that when opened allows a blast of steam to discharge any material in the drum.

Nonconducting material: Any material used as insulation around the boiler shell and connecting steam pipes. In the Civil War era, firebricks were always used around the furnaces, whereas other materials such as asbestos, clay, and mixtures of earth and straw were used for insulating the boiler tops.

Shell plate: Metal plating, usually cast iron, composing the outside walls of a boiler.

Smokestack (chimney, smoke pipe, funnel): A large tube allowing the passage of exhaust gases to the atmosphere. It is made tall to encourage natural draft and to protect topsides from soot and cinders.

Steam drum: A hollow metal cylinder, usually mounted transversely above the after end of a boiler set, used for collecting steam for transmission to the engines via the main steam line.

Watertube: A boiler type in which exhaust gases surround many small water-filled tubes to heat the water inside.

Tube plate: Metal plating, usually cast iron or brass, composing the walls of the boiler firetubes or watertubes.

ENGINES AND PROPULSION

Air pump (vacuum pump): A pump used for maintaining a vacuum and flushing out air, injection water, or condensate in the condenser, keeping it operating at maximum efficiency.

Bedplate: A single cast-metal plating on which the engines or boilers are mounted to a vessel's floor timbers and/or engine mounting blocks.

Cam: The generic term for any device that converts rotary motion to oscillating motion.

Connecting rod: A metal or wooden shaft linked to the piston rod at one end and the main shaft (propeller or paddlewheel) at the other. It converts the oscillating motion of the piston to semirotary motion.

Crank: A wood or metal fitting for converting the semirotary motion of the connecting rod to rotary motion for turning the propeller or paddlewheel.

Crank pin: A pin used to link the connecting rod to the crank; it provides a pivot point for converting semirotary motion to rotary motion for turning the propeller or paddlewheel.

Crankpit: Space cut out beneath the engine mounts to provide room for the rotation of the engine-connecting rods and eccentrics. It is one of the lowest points inside a steamer's hull.

Crank shaft: Any shaft connected to a crank; it usually denotes a connecting rod.

Crosshead: A metal slide or guide for supporting the linkage between the piston rod and the connecting rod. It converts the oscillating motion of the piston to semirotary motion for the connecting rod.

Cylinder: A large thick-walled metal tube; the primary engine component. It contains and houses the piston, which generates power when pushed by admitted steam. Dimensional measurements can be given for the outside or inside (bore) diameter. The length is given in stroke (length of piston thrust).

Cylinder head: A thick metal plate covering the end of the cylinder. It is bolted down but is removable if access to the internal parts is necessary.

Delivery pipe (outboard or inboard): An outboard delivery pipe discharges overboard any excess water not needed for the boilers; an inboard delivery pipe takes on any extra water that is needed. Discharge and intake occur continuously as part of the normal circulation in a condensing system.

Eccentric: An off-centered disc coupled to a shaft; it is a type of cam. Eccentrics are located primarily on the main drive shaft for the operation of the reversing gear through a linkage to the rock-shaft arm and thus the valve control.

Exhaust valve: A valve that allows the escape of steam from the cylinder after use.

Expansion valve: A valve that improves the efficiency of the engine by closing off the steam supply before the piston has traveled a full stroke. This is called *cutoff.* Any steam already in the cylinder expands to finish driving the piston, so an expansion valve allows less steam to be used and more fuel to be saved.

Follower: A metal flange bolted to the backside or underside of the piston to secure the packing and also to provide a surface for steam to act against if the engine is double acting.

Idler shaft: Any shafting not directly contributing to propulsion; it is usually a rotating connection between two separate propeller shafts that prevents the independent turning of each screw.

Levers and wipers: These are most common on western river–style engines. A centrally placed rock shaft moves a pair of wing-shaped wipers that alternately lift or drop levers connected to the steam and exhaust valves on either end of the engine cylinder.

Link: The connection between moving parts, most often between valve gear components.

Main shaft (drive shaft): A large metal rod imparted with rotary motion by engine linkages, to which is fixed a vessel's primary means of propulsion: screws or paddlewheels.

Main valve: The primary valve controlling the direction of steam entry to the engine cylinder; operating it engages the valve or reversing gear to change the direction of the engine.

Packing: A ring of material—most commonly hemp, but many other materials have been used—around the piston and valve rods, where they pass through points that must be sealed, such as the cylinder ends. Improper or infrequent repacking has been the cause of many troubles.

Pillow block: A cast-metal piece that supports and contains the bearing components of the main shaft.

Piston: The disc-shaped moving part of an engine on which the pressure of steam acts to provide power. Some engines are single acting, with steam pressure driving the piston only one way, but most are double acting, with steam admitted and exhausted at both ends of the engine cylinder, driving the piston back and forth.

Piston head: The disc-shaped part of the piston on which the pressure of steam acts to impart motion.

Pittman: A long wooden arm used on river steamboats to connect the piston rod to the paddlewheel crank—basically a very large and long connecting rod.

Poppet valve: A simple type of valve that operates like a cork in a bottle.

Reversing gear (valve gear): The linkage of levers and arms to cams or eccentrics that allows the reversal of the direction of steam into the cylinders, which reverses the engine. There are several different types, but most common is the Stephenson valve gear, originally developed for steam locomotives.

Ring: A circular seal that fits between the piston head and the cylinder wall.

Rock-shaft arm: A cast-metal arm moving in a rocking motion from a central pivot or bearing, used to alternately open and close the steam and exhaust valves on the engine cylinder.

Screw: A common technical name for the propeller, describing its form.

Starting valve: A small auxiliary valve admitting steam directly from the boilers into one end of one cylinder to assist in starting the engines.

Steam chest (valve chest, valve chamber): The casing that houses an engine's main valves; it lets steam into the cylinder as well as exhaust for its removal. The time that both valves are closed to allow for compression is termed *lap*. The steam chest is usually attached to the top of the cylinder, more rarely on the side, making it easily accessible for maintenance.

Stuffing box and stuffing gland: A stuffing box is a chamber for packing that is covered and secured by a fitted metal plate called a stuffing gland. This setup is used anywhere that a seal around moving parts is required, mostly around piston rods and the main drive shaft, where it passes through the hull.

Telescopic joint (expansion joint): A loop-shaped extension in a steam line to account for the expansion and contraction of the pipes from changing temperatures. It also provides some flexibility for long steam pipes in a riverboat's flexible hull.

Throttle: A device that controls steam flow rate, actuated by a wheel or a lever control to a valve in the main steam line. It is operated by the chief engineer or the highest-ranking engineer on duty. The engineer's primary post is at the throttle platform.

Thrust bearing: The main bearing on the propeller shaft, designed to prevent forward motion of the shaft *independent of the hull* from the propeller's driving force. A series of strong hardened metal interlocking collars fit around the shaft at a bearing and are very firmly secured to the framing of the vessel.

Wheel (paddlewheel or propeller): A device for moving a steam vessel. The term is mostly applied to the paddlewheel but is sometimes used to denote a propeller. It should not be confused with a ship's steering wheel when used in an engineering context.

Wrist pin: A pin used to link the piston rod to the connecting rod. It provides a pivot point for converting the oscillating motion of the former to the semirotary motion of the latter.

AUXILIARY EQUIPMENT

Capstan: A deck-mounted cylinder rotating on a vertical axis used for winding cable or rope. It can be powered or turned manually using levers.

Condenser: A large piece of equipment used to condense exhaust steam from the engines into water for recirculation in the boilers. Two main types are the jet and the closed (or surface). The former mixes a jet of cold water with exhaust steam; the latter works by surrounding exhaust steam with cold water tubes. Vessels operating on freshwater often did not have condensers; machinery without this component is termed *noncondensing*.

Doctor: A small engine, usually a beam type, equipped with a feedwater heater and used for various pumping duties.

Donkey: A small, usually vertical, boiler and/or its attendant engine, usually a small horizontal type. It is used to generate power for capstans and fan blowers, among other things.

Fan blower (forced draft fan): A fan placed in the entrance to the smokestack, powered by a donkey boiler and engine, used for inducing an artificial or forced draft for the boiler fires.

Feedwater heater: A metal cylinder filled with steam tubes for heating the water that returns to the boilers.

Gauge cock (try cock): A nozzle with a small valve wheel used for checking the boiler water level. It is usually arranged in a column of three: when the boiler water is at the ideal level, the top, when opened, should emit steam; the middle should emit a water-steam mixture; and the bottom should emit water only.

Hot well: A small reservoir and filter used to collect condensed steam in preparation for its return to the boiler as feedwater.

Hydrostatic lubricator: A small device fitted for supplying lubricating oil through doses regulated by water displacement.

Injector: A device used for propelling feedwater into the boiler using a steam jet. It is often called Giffard's injector, after the inventor, or the live steam injector, for its use of steam directly from the boiler.

Lever safety valve: A valve fitted to the top of the boilers for emitting excess steam and preventing dangerous boiler overpressure. The pressure at which steam is emitted can be regulated by moving a small weight (called a *pea*) at the end of a lever connected to a valve mechanism.

Main force pump: A ship's hand-powered main bilge pump. It is used as a backup for all other pumps in case of breakdown and for filling the boilers, making tests, and emptying the bilges.

Pump well: Technically (and ideally), the lowest point inside a vessel into which all bilgewater and other materials flow for easy removal. The main pumps have their suction here.

Worthington pump: A powerful horizontal pump with its own attached engine, used for a variety of purposes and common to many vessels. Two types were made: simplex (single cylinder, single pump) and duplex (two cylinders and pumps). It is named after its inventor.

BIBLIOGRAPHY

Alexander, Violet. "The Confederate States Navy Yard at Charlotte, N.C., 1862–1865." *Southern Historical Society Papers* 40, no. 2 (1915): 184–93.

Amadon, George F. *Rise of the Ironclads.* Missoula, MT: Pictorial Histories, 1988.

"Archaeological Research at West Point Foundry Preserve." Scenic Hudson, http://web.archive.org/web/20070629101949/http://www.scenichudson.org/land_pres/wpfp_research.htm#history.

Atherton, Charles. *On Marine Engine Construction and Classification.* London: John Weale Architectural Library, 1851.

Atlantic Works v. Brady. 107 US 192 (1883). Open Jurist, http://openjurist.org/107/us/192/atlantio-works-v-brady-brady.

Ball, David A. "*Phoenix,* and the Confederate Obstructions of Upper Mobile Bay, Alabama." Master's thesis, Florida State University, Tallahassee, 1998.

Bates, Alan. *The Western Rivers Engineroom Cyclopoedium.* Louisville, KY: Cyclopoedium Press, 1996.

———. *The Western Rivers Steamboat Cyclopoedium, or American Riverboat Structure and Detail, Salted with Lore, with a Nod to the Modelmaker.* Leonia, NJ: Hustle Press, 1968.

Bathe, Granville. *Ship of Destiny: A Record of the U.S. Steam Frigate Merrimac, 1855–1862.* St. Augustine, FL: n.p., 1951.

Baxter, James Phinney III. *The Introduction of the Ironclad Warship.* Cambridge, MA: Harvard University Press, 1933.

Bennett, Frank M. *The Steam Navy of the United States: A History of the Growth of the Steam Vessel of War in the U.S. Navy, and of the Naval Engineer Corps.* Pittsburgh, PA: W. T. Nicholson, 1896. https://archive.org/details/cu31924030755361.

Bishop, J. Leander. *A History of American Manufacturers from 1608 to 1860.* Philadelphia, PA: Edward Young, 1864. https://play.google.com/store/books/details?id=K1YdAAAAIAAJ.

Bright, Leslie S., William H. Rowland, and James C. Bardon. *CSS Neuse: A Question of Iron and Time.* Raleigh: North Carolina Department of Cultural Resources, 1981.

Brooke, George M. Jr., ed. *Ironclads and Big Guns of the Confederacy: The Journal and Letters of John M. Brooke.* Columbia: University of South Carolina Press, 2002.

Brown, David K. *Paddle Warships: The Earliest Steam-Powered Fighting Ships, 1815–1850.* London: Conway Maritime Press, 1993.

Bruce, Kathleen. *Virginia Iron Manufacture in the Slave Era.* New York: Century, 1931.

Buchanan, Franklin. Letterbook. Buchanan and Screven Family Papers, 1801–1913. Southern Historical Collection, University of North Carolina, Chapel Hill.

"The Buffalo Dry Dock." *The S.S. Canadiana History Book*. SS Canadiana, http://sscanadiana.com/buffdrydock.html.

Bureau of Ships. Records, 1794–1972. Record Group 19. National Archives and Records Administration, College Park, MD.

"Burned." *Wilmington (NC) Morning Star*, September 8, 1871.

Campbell, Peter B. "The Development of Confederate Ship Construction: An Archaeological and Historical Investigation of Confederate Ironclads *Neuse* and *Jackson*." Master's thesis, East Carolina University, Greenville, NC, 2009.

Campbell, R. Thomas. *Engineer in Gray: Memoirs of Chief Engineer James H. Tomb, CSN*. Jefferson, NC: McFarland, 2005.

———. "The First Confederate Ironclad." *America's Civil War*, July 2006, 46–52.

———. *Midshipman in Gray: Selections from Recollections of a Rebel Reefer*. Shippensburg, PA: Burd Street Press, 1997.

———. *Southern Service on Land and Sea: The Wartime Journal of Robert Watson, CSA/CSN*. Knoxville: University of Tennessee Press, 2002.

Canney, Donald L. *The Old Steam Navy: Frigates, Sloops, and Gunboats, 1815–1885*. Annapolis, MD: US Naval Institute Press, 1990.

Carbone, John Stephen. *The Civil War in Coastal North Carolina*. Raleigh, NC: Department of Cultural Resources, 2001.

Coker, P. C. III. *Charleston's Maritime Heritage, 1670–1865*. Charleston, SC: CokerCraft Press, 1987.

Combs, Edwin L. III. "On Duty at Wilmington: The Confederate Navy on the Cape Fear River." Master's thesis, East Carolina University, Greenville, NC, 1996.

Continental Eagle Corporation. "Factory at War." Pratt History, http://www.pratthistory.com/factory_at_war.htm.

Coski, John M. *Capital Navy: The Men, Ships and Operations of the James River Squadron*. New York: Savas Beatie, 2005.

Daniel, Larry J., and Riley W. Gunter. *Confederate Cannon Foundries*. Union City, TN: Pioneer Press, 1977.

Dent, John Horry. Papers, 1840–1892. W. S. Hoole Special Collections Library, University of Alabama, Tuscaloosa.

Dew, Charles B. *Ironmaker to the Confederacy: Joseph R. Anderson and the Tredegar Iron Works*. Wilmington, NC: Broadfoot, 1987.

Donnelly, Ralph W. "The Charlotte, North Carolina, Navy Yard, CSN." *Civil War History* 5 (1959): 72–79.

Driver, Robert J. Jr. *Confederate Sailors, Marines, and Signalmen from Virginia and Maryland*. Westminster, MD: Heritage Books, 2007.

Duffield, Ulysses Grant. *The Neafie and Levy Ship and Engine Building Company (Penn Works)*. Philadelphia, PA: Franklin, 1896.

Durkin, Joseph T. *Confederate Navy Chief: Stephen R. Mallory*. Tuscaloosa: University of Alabama Press, 1954.

Eldridge, Elwin. Collection on John H. Morrison, 1881–1921. MS0006. Mariners' Museum Library, Christopher Newport University, Newport News, VA.

Elliott, Robert G. *Ironclad of the Roanoke: Gilbert Elliott's Albemarle*. Shippensburg, PA: White Mane, 2005.

Flanders, Alan B. *John L. Porter: Naval Constructor of Destiny*. White Stone, VA: Brandylane, 2000.

Flanders, Alan B., and Neale O. Westfall, eds. *Memoirs of E. A. Jack: Steam Engineer, CSS Virginia*. White Stone, VA: Brandylane, 1998.

Foster, Kevin J. "Marine Steam Engines, 1807–1875: A Typology." In *Archaeology in Solution: Proceedings of the Seventeenth Annual Conference on Underwater Archaeology*, edited by John W. Foster and Sheli O. Smith, 67–70. Salinas, CA: Fathom Eight, 1986.

Freedley, Edwin T., ed. *United States Mercantile Guide: Leading Pursuits and Leading Men, a Treatise on the Principal Trades and Manufactures of the United States*. Philadelphia, PA: Edward Young, 1856. https://play.google.com/store/books/details?id=qvcYAAAAYAAJ.

Gift, George Washington. Letters, 1862–1876. Southern Historical Collection, University of North Carolina, Chapel Hill.

Gift, George W. Papers, 1862–1876. Collection 01152. Southern Historical Collection, University of North Carolina, Chapel Hill. http://www2.lib.unc.edu/mss/inv/g/Gift,George_W.html.

Gosnell, H. Allen. *Guns on the Western Waters: The Story of River Gunboats in the Civil War*. Baton Rouge: Louisiana State University Press, 1949.

Gulf Engineers and Consultants and Tidewater Atlantic Research. *Archeological Documentation and Testing of a Mid-Nineteenth Century Tugboat at Hutchinson Island, Savannah, Chatham County, Georgia: Final Report*. Savannah, GA: US Army Corps of Engineers, 1992.

Gutierrez, Joseph A. Jr. "Confederate Naval Ordnance." Master's thesis, East Carolina University, Greenville, NC, 1977.

Harris, Lynn. "Wreck of the Golden Spike." *Legacy* 8, no. 1 (2003): 8–9.

Hearn, Chester G. *Mobile Bay and the Mobile Campaign*. Jefferson, NC: McFarland, 1993.

Hewett, Janet B., ed. *Supplement to the Official Records of the Union and Confederate Armies*. Vol. 3. Wilmington, NC: Broadfoot, 1999.

Heyl, Erik. *Early American Steamers*. Buffalo, NY: Erik Heyl, 1953.

Hillery, Rev. Horace S. Papers, 1953–1962. HC404. Historical Records Series, Historian's Collection, Putnam County Historian, Brewster, NY.

Holcombe, A. Robert Jr. "CSS *Savannah*." *Warship International* 20, no. 3 (1983): 309–14.

———. "The Evolution of Confederate Ironclad Design." Master's thesis, East Carolina University, Greenville, NC, 1993.

———. "The *Richmond* Class Confederate Ironclads." *CHAB (Confederate Historical Association of Belgium) News* 6 (1978): 16–20.

"Ingomar." *Louisville (KY) Daily Courier*, January 8, 1855.

"Iron-Clad Ships of War." *Blackwood's Edinburgh* 88, no. 537 (1860): 616–49.

Jeter, Katherine Brash. "Against All Odds: Lt. Jonathan H. Carter, CSN, and His Ironclad." *Louisiana History* 28, no. 3 (1987): 263–88.

———. *A Man and His Boat: The Civil War Career and Correspondence of Lieutenant Jonathan H. Carter, CSN*. Lafayette: Center for Louisiana Studies, 1996.

Johnson, Robert U., and Clarence C. Buell, eds. *Battles and Leaders of the Civil War*. 4 vols. 1887–1888. Reprint, Harrisburg, PA: Archive Society, 1991.

Johnston, J. D. "Admiral Buchanan and the Confederate States Ram 'Tennessee.'" *United Service* 7, no. 2 (1882): 202–13.

Kane, Adam Isaac. "The Western River Steamboat: Structure and Machinery, 1811 to 1860." Master's thesis, Texas A&M University, College Station, 2001.

Kean, P. *Report of Evidence Taken before a Joint Special Committee of Both Houses of the Confederate Congress to Investigate the Affairs of the Navy Department*. Richmond, VA: Geo. P. Evans, 1863.

Kleber, John E., ed. *The Encyclopedia of Louisville*. Lexington: University Press of Kentucky, 2001.

Lamar, G. B., John Stoddard, and Francis Sorrel. "The Floating Battery *Georgia*." *Savannah (GA) Republican*, October 21, 1862.

Lambert, Andrew, ed. *Steam, Steel, and Shellfire: The Steam Warship, 1815–1905*. Edison, NJ: Conway Maritime Press, 1992.

Lander, Ernest M. Jr. "Charleston: Manufacturing Center of the Old South." *Journal of Southern History* 26, no. 3 (1960): 330–51.

Lang, S. Ruby. "Research on the Confederate Ironclad Ram, Jackson." In *Underwater Archaeology: The Challenge before Us; The Proceedings of the Twelfth Conference on Underwater Archaeology*, edited by Gordon P. Watts, 64–68. San Marino, CA: Fathom Eight, 1981.

Lankford, Nelson D., ed. *An Irishman in Dixie: Thomas Conolly's Diary of the Fall of the Confederacy*. Columbia: University of South Carolina Press, 1988.

"Leeds Iron Foundry—New Orleans, LA." Waymarking. http://www.waymarking.com/waymarks/WM14QA_Leeds_Iron_Foundry_New Orleans_LA.

Library of Congress. *Prints and Photographs Online Catalog*. 2016. http://www.loc.gov/pictures/.

Loring, Harrison. Collection. Collection 226. Manuscript Collection Registers, Collections Research Center, Mystic Seaport, Mystic, CT. http://library.mysticseaport.org/manuscripts/coll/coll226.cfm.

Luraghi, Raimondo. *A History of the Confederate Navy*. Translated by Paolo E. Coletta. Annapolis, MD: US Naval Institute Press, 1996.

Lytle, William M., and Forrest R. Holdcamper. *Merchant Steam Vessels of the United States, 1790–1868*. Staten Island, NY: Steamship Historical Society of America, 1975.

MacBride, Robert. *Civil War Ironclads: The Dawn of Naval Armor*. Philadelphia, PA: Chilton Books, 1962.

Maffitt, John N. "Reminiscences of the Confederate Navy." *United Service* 3, no. 1 (1880): 495–514.

Martin, William Francis. Papers, 1787–1884. Southern Historical Collection, University of North Carolina, Chapel Hill.

McBlair, C. H. "A Historical Sketch of the Confederate Navy." *United Service* 3, no. 1 (1880): 588–613.

McClinton, Oliver Wood. "The Career of the Confederate States Ram 'Arkansas.'" *Arkansas Historical Quarterly* 7, no. 4 (1948): 329–33.

McLaughlin, Augustus. Letterbook, 1864–1865. Port Columbus National Civil War Naval Museum, Columbus, GA.

Melton, Maurice. *The Best Station of Them All: The Savannah Squadron, 1861–1865*. Tuscaloosa: University of Alabama Press, 2012.

———. *The Confederate Ironclads*. New York: Thomas Yoseloff, 1968.

Merrill, James M. "Confederate Shipbuilding at New Orleans." *Journal of Southern History* 28, no. 1 (1962): 87–93.

Miller, Francis T., and Robert S. Lanier, eds. *The Photographic History of the Civil War in Ten Volumes*. New York: Review of Reviews, 1911. https://archive.org/stream/photographichisto1inmill#page/n9/mode/2up.

Minor Family Papers, 1810–1932. Virginia Historical Society, Richmond, VA.

Mooney, James L., ed. *Dictionary of American Naval Fighting Ships*. 8 vols. Washington, DC: US Department of the Navy, 1959–1981.

Mueller, Edward A. *Perilous Journeys: A History of Steamboating on the Chattahoochee, Apalachicola, and Flint Rivers, 1828–1928*. Eufaula, AL: Historic Chattahoochee Commission, 1990.

Neville, Bert. *Directory of Steamboats with Illustrations and Lists of Landings on the Chattahoochee-Apalachicola-Flint-Chipola Rivers*. Selma, AL: Coffee, 1961.

"New Towboat." *Boston Evening Transcript*, December 5, 1854.

Office of Naval Records and Library (ONRL). Naval Records Collection, 1691–1945, Record Group 45. National Archives and Records Administration, College Park, MD.

"Our Boat Building . . . " *Louisville (KY) Daily Courier*, January 19, 1857.

Panamerican Consultants and Tidewater Atlantic Research. *In Situ Archaeological Evaluation of the CSS Georgia, Savannah Harbor, Georgia: Final Report*. Savannah, GA: US Army Corps of Engineers, 2007.

Park, Carl D. *Ironclad Down: The USS Merrimack–CSS Virginia from Construction to Destruction*. Annapolis, MD: US Naval Institute Press, 2007.

Parker, Larry. "For Want of a Nail: An Evaluation of the Confederate Ironclad's Construction History, Service History, Tactical and Strategic Employment." Military History Online, http://www.militaryhistoryonline.com/civilwar/articles/wantofnail.aspx.

Parker, William H. *Recollections of a Naval Officer, 1841–1865*. New York: Charles Scribner's Sons, 1883.

Pearce, George F. *Pensacola during the Civil War: A Thorn in the Side of the Confederacy*. Gainesville: University Press of Florida, 2000.

Peebles, Martin D. "CSS *Raleigh*: The History and Archaeology of a Civil War Ironclad in the Cape Fear River." Master's thesis, East Carolina University, Greenville, NC, 1996.

Peek, Charles Smith. Papers, 1863–1865. Collection of Dr. Charles Peery, Charleston, SC.

Poalk, W. L. "Advertisement." *North Carolina Daily Times* (New Bern, NC), October 3, 1865.

"Propeller Enoch Train, of Boston." *U.S. Nautical Magazine and Naval Journal* 3 (1855–1856): 221.

Ragan, William. "The *Louisiana*." Unpublished manuscript, 2012.

———. "The *Mississippi*." Unpublished manuscript, 2011.

Reanie, Neafie & Company Papers, 1844–1861. MS102. Mariners' Museum Library, Christopher Newport University, Newport News, VA.

Reaves, Bill. *History of Southport*. Southport, NC: Southport Historical Society, 1978.

"The Rebel War Vessels." *Boston Daily Journal*, August 9, 1861.

Ripley, Warren. *Artillery and Ammunition of the Civil War*. 4th ed. Charleston, SC: Battery Press, 1984.

Roper, Stephen. *Handbook of Land and Marine Engines, including the Modeling, Construction, Running, and Management of Land and Marine Engines and Boilers*. 9th ed. Philadelphia, PA: David McKay, 1897.

Roy, John. Diary, 1860–1862. Howard-Tilton Memorial Library, Tulane University, New Orleans, LA.

Sagle, Lawrence W., and Alvin F. Staufer. *B&O Power: Steam, Diesel, and Electric Power of the Baltimore and Ohio Railroad, 1829–1964*. N.p.: A. F. Staufer, 1964.

Saltus, Allen R., and Sidney H. Schell. *The CSS Huntsville and CSS Tuscaloosa Project, Mobile River, Alabama: Report of 1985 Activities*. Mobile, AL: US Army Corps of Engineers and Mobile Jaycees, 1985.

Savannah Squadron Papers, 1862–1865. Manuscript, Archives, and Rare Book Library, Robert W. Woodruff Library, Emory University, Atlanta, GA.

Scharf, J. Thomas. *History of the Confederate States Navy from Its Organization to the Surrender of Its Last Vessel*. 1886. Reprint, New York: Crown, 1977.

Sheppard, Robert M. III. "U.S.S. *Cairo* Engines and Boilers." American Society of Mechanical Engineers, 1990. https://www.asme.org/getmedia/2306f17b-2b7f-4592-bea8-073cf64700fa/143-USS-Cairo-Engine-and-Boilers-1862.aspx.

Silverstone, Paul H. *Civil War Navies, 1855–1883*. Annapolis, MD: US Naval Institute Press, 2001.

———. *Warships of the Civil War Navies*. Annapolis, MD: US Naval Institute Press, 1989.

Smith, Edgar C. *A Short History of Naval and Marine Engineering*. Cambridge, UK: Cambridge University Press, 1937.

Smith, Myron J. Jr. *The CSS Arkansas: A Civil War Ironclad on Western Waters*. Jefferson, NC: McFarland, 2011.

"Some Fifty Tons of Iron Removed from Cape Fear." *Wilmington (NC) Morning Star*, July 1, 1868.

Sprunt, James. *Chronicles of the Cape Fear River, 1660–1916*. Raleigh, NC: Edwards and Broughton, 1916.

Standard, Diffie W. *Columbus, Georgia, in the Confederacy*. New York: William-Frederick Press, 1954.

"Steamer Baltic." *New Albany (IN) Daily Ledger*, December 16, 1856.

Stickney, John M. *Promotion or the Bottom of the River: The Blue and Gray Naval Careers of Alexander F. Warley, South Carolinian.* Columbia: University of South Carolina Press, 2012.

Still, William N., ed. *The Confederate Navy: The Ships, Men and Organization, 1861–65.* Annapolis, MD: US Naval Institute Press, 1997.

———. *Confederate Shipbuilding.* Athens: University of Georgia Press, 1987.

———. "Facilities for the Construction of War Vessels in the Confederacy." *Journal of Southern History* 31, no. 3 (1965): 285–304.

———. *Iron Afloat: The Story of the Confederate Armorclads.* Columbia: University of South Carolina Press, 1985.

———. *The Savannah Squadron.* Savannah, GA: Coastal Heritage Press, 1989.

Swanson, Mark, and Robert Holcombe. *CSS Georgia: Archival Study.* Stone Mountain, GA: New South Associates, 2007.

Taggart, Lewis Williams. "Technology and the Confederate Navy." Master's thesis, University of Tulsa, OK, 1991.

Talbott & Brother. *An Illustrated Catalogue of Machinery, Manufactured by Talbott & Brother, Richmond, VA.* Richmond, VA: Chas. H. Wynne, Printer, 1860.

Talbott & Brother Papers, 1831–1880. Virginia Historical Society, Richmond, VA.

Tellon, Mart. "In the Heyday of Steamboating in New Albany." *New Albany (IN) Weekly Ledger,* May 28, 1924. http://webcache.googleusercontent.com/search?q=cache:http://www.nafclibrary.org/wp-content/uploads/2013/08/HeydayofSteamboating.pdf.

Thiesen, William H. *Industrializing American Shipbuilding: The Transformation of Ship Design and Construction, 1820–1920.* Gainesville: University Press of Florida, 2006.

Tomblin, Barbara B. "From Sail to Steam: The Development of Steam Technology in the United States Navy 1838–1865." PhD dissertation, Rutgers University, New Brunswick, NJ, 1988.

"The Tow-Boat 'Baltic.'" *New Albany (IN) Tribune,* December 10, 1856.

"Towboat *Enoch Train.*" *Boston Daily Journal,* March 13, 1855.

Trotter, William R. *Ironclads and Columbiads: The Civil War in North Carolina, the Coast.* Winston-Salem, NC: John F. Blair, 1989.

Turner, Maxine. *Navy Gray: A Story of the Confederate Navy on the Chattahoochee and Apalachicola Rivers.* Tuscaloosa: University of Alabama Press, 1988.

"The T. W. Roberts." *Louisville Daily Courier,* November 19, 1860.

Underwood, Rodman L. *Stephen Russell Mallory: A Biography of the Confederate Navy Secretary and United States Senator.* Jefferson, NC: McFarland, 2005.

US Department of the Navy. *Dictionary of American Naval Fighting Ships.* Naval History and Heritage Command, https://www.history.navy.mil/research/histories/ship-histories/danfs.html.

———. *Official Records of the Union and Confederate Navies in the War of the Rebellion.* Series 1, 27 vols.; series 2, 3 vols. 1894–1922. Reprint, Harrisburg, PA: National Historical Society, 1987.

———. Photo Archive. Naval History and Heritage Command, http://www.history.navy.mil/our-collections/photography.html.

US Department of the Treasury. Confederate Records, 1833–1878. Record Group 365. National Archives and Records Administration, College Park, MD.

———. General Records, 1789–1990. Record Group 56. National Archives and Records Administration, College Park, MD.

US Department of War. Confederate Records, 1825–1900. Record Group 109. National Archives and Records Administration, College Park, MD.

———. *The War of the Rebellion: A Compilation of the Official Records of the Union and Confederate Armies.* Series 1, 53 vols.; series 2, 8 vols.; series 3, 5 vols; series 4; 3 vols. 1880–1901. Reprint, Harrisburg, PA: National Historical Society, 1985.

US Light House Establishment. *Manual of Light-House Engineering, Embracing Plans of Towers, Light Vessels, Beacons, Bouys, and Other Aids to Navigation, Designed and Executed under the Direction of the Light House Board, and Exhibiting the Present System of Lenses, Lamps, Implements.* Washington, DC: Light-House Board, 1869.

Warner, James. Letterbook, 1862–1865. Port Columbus National Civil War Naval Museum, Columbus, GA.

Watts, Gordon P. *Underwater Archaeological Documentation of the Remains of the Ironclad CSS North Carolina, Brunswick County, North Carolina.* Wilmington, NC: US Army Engineer District, 1999.

Way, Frederick Jr. *Way's Packet Directory, 1848–1994.* Athens: Ohio University Press, 1983.

Wells, Tom Henderson. *The Confederate Navy: A Study in Organization.* Tuscaloosa: University of Alabama Press, 1971.

Wexler, Charles James. "'The Ordinary Gunboats': The CSS *Chicora*, the CSS *Palmetto State*, and the Battle off Charleston Harbor, January 31, 1863." Master's thesis, Citadel, Charleston, SC, 2008.

Wilson, Clyde E. Gunboat *Baltic* Papers, 1938. Public Information Subject Files: Navy, SG11146. Alabama Department of Archives and History, Montgomery.

Wilson, H. W. *Ironclads in Action: A Sketch of Naval Warfare from 1855 to 1895, with Some Account of the Development of the Battleship in England.* 4th ed. Vol. 1. London: Sampson Low, Marston, 1896. https://play.google.com/store/books/details?id=RHDVAAAAMAAJ.

Work Projects Administration. *Ship Registers and Enrollments of New Orleans, Louisiana.* Vol. 5, *1851–1860.* Baton Rouge: Louisiana State University, 1942. http://www.youngsanders.org/shipvol5.pdf.

Wright, Henry X. Papers, 1859–1874. MS0072. Mariners' Museum Library, Christopher Newport University, Newport News, VA.

"Y-Z, C.S. Navy Sailors." *NC Civil War Sailors Project: Identifying NC Sailors and Marines.* http://rblong.net/sailor/y-z.html.

ILLUSTRATION CREDITS

Figure 1.1. A. Robert Holcombe Jr., "The Evolution of Confederate Ironclad Design" (master's thesis, East Carolina University, Greenville, NC, 1993), 149–50.

Figure 1.2. Ibid., 99.

Figure 1.3. US Department of the Navy, "NH 55226 CSS *Chattahoochee* (1863–1864)," Naval History and Heritage Command Photo Archives, http://www.history.navy.mil/our-collections/photography/numerical-list-of-images/nhhc-series/nh-series/NH-55000/NH-55226.html.

Figure 1.4. US Department of the Navy, "NH 41613 US Navy 88-Foot Harbor Tug," Naval History and Heritage Command Photo Archives, http://www.history.navy.mil/our-collections/photography/numerical-list-of-images/nhhc-series/nh-series/NH-41000/NH-41613.html.

Figure 1.5. Alan Bates, *The Western Rivers Engineroom Cyclopoedium* (Louisville, KY: Cyclopoedium Press, 1996), 41–42.

Figure 1.6. Edgar C. Smith, *A Short History of Naval and Marine Engineering* (Cambridge, UK: Cambridge University Press, 1937), 146.

Figure 1.7. Bates, *Western Rivers Engineroom*, 8.

Figure 1.8. Courtesy of A. Robert Holcombe Jr., 2012.

Figure 1.9. Granville Bathe, *Ship of Destiny: A Record of the U.S. Steam Frigate Merrimac, 1855–1862* (Philadelphia, PA: Allen, Lane, and Scott, 1951), plate 2.

Figure 1.10. Bates, *Western Rivers Engineroom*, 20–21.

Figure 1.11. Ibid., 16.

Figure 1.12. Ibid., 14.

Figure 1.13. Ibid., 17, 19, 28.

Figure 1.14. Ibid., 47.

Figure 2.1. Dirk Soulis Auctions, "19th C. American Maritime Painting of the Enoch Train," lot 66, Live Auctioneers, https://www.liveauctioneers.com/item/27408045_19th-c-american-maritime-painting-of-the-enoch-train.

Figure 2.2. Andrew Lambert, ed., *Steam, Steel, and Shellfire: The Steam Warship, 1815–1905* (Edison, NJ: Conway Maritime Press, 1992), 174.

Figure 2.3. US Department of the Navy, "NH 79598 CSS Manassas (1861–1862)," Naval History and Heritage Command Photo Archives, http://www.history.navy.mil/our-collections/photography/numerical-list-of-images/nhhc-series/nh-series/NH-79000/NH-79598.html.

Figure 2.4. Donald L. Canney, *The Old Steam Navy: Frigates, Sloops, and Gunboats, 1815–1885* (Annapolis, MD: US Naval Institute Press, 1990), 47.

Figure 2.5. Carl D. Park, *Ironclad Down: The USS Merrimack–CSS Virginia from Construction to Destruction* (Annapolis, MD: US Naval Institute Press, 2007), 74.

Figure 2.6. US Department of the Navy, "NH 58793 CSS *Baltic* (1862–1864)," Naval History and Heritage Command Photo Archives, http://www.history.navy.mil/our-collections/photography/numerical-list-of-images/nhhc-series/nh-series/NH-58000/NH-58793.html.

Figure 3.1. Robert U. Johnson and Clarence C. Buell, eds., *Battles and Leaders of the Civil War* (1887, repr., Harrisburg, PA: Archive Society, 1991), 2:49.

Figure 3.2. US Department of the Navy, "NH 57272 CSS *Arkansas* (1862–1862)[*sic*]," Naval History and Heritage Command Photo Archives, http://www.history.navy.mil/our-collections/photography/numerical-list-of-images/nhhc-series/nh-series/NH-57000/NH-57272.html.

Figure 3.3. US Department of the Navy, "NH 58721 The Rebel Ironclad 'Georgia,'" Naval History and Heritage Command Photo Archives, https://www.history.navy.mil/our-collections/photography/numerical-list-of-images/nhhc-series/nh-series/NH-58000/NH-58721.html.

Figure 4.1. US Light House Establishment, *Manual of Light-House Engineering, Embracing Plans of Towers, Light Vessels, Beacons, Buoys, and Other Aids to Navigation, Designed and Executed under the Direction of the Light House Board, and Exhibiting the Present System of Lenses, Lamps, Implements* (Washington, DC: Light House Board, 1869), plate 52). Courtesy of Bil Ragan, 2014.

Figure 4.2. Courtesy of A. Robert Holcombe Jr., 2012.

Figure 4.3. James Barnes, *The Navies*, vol. 6 of *The Photographic History of the Civil War in Ten Volumes*, ed. Francis T. Miller and Robert S. Lanier (New York: Review of Reviews, 1911), 239, https://archive.org/stream/photographichisto1inmill#page/n9/mode/2up.

Figure 4.4. William N. Still Jr., ed., *The Confederate Navy: The Ships, Men and Organization, 1861–65* (Annapolis, MD: US Naval Institute Press, 1997), 44.

Figure 4.5. Courtesy of Holcombe, 2012.

Figure 4.6. William N. Still Jr., *Iron Afloat: The Story of the Confederate Armorclads* (Columbia: University of South Carolina Press, 1985), 103.

Figure 5.1. US Department of the Navy, "NH 60335 USS *Tennessee* (1864–1867)," Naval History and Heritage Command Photo Archives, http://www.history.navy.mil/our-collections/photography/numerical-list-of-images/nhhc-series/nh-series/NH-60000/NH-60335.html.

Figure 5.2. Courtesy of Holcombe, 2012.

Figure 5.3. William N. Still Jr., *Confederate Shipbuilding* (Athens: University of Georgia Press, 1987), 60.

Figure 6.1. US Department of the Navy, "NH 76391 Confederate 180-

Foot Ironclad Gunboat," Naval History and Heritage Command Photo Archives, http://www.history.navy.mil/our-collections/photography/numerical-list-of-images/nhhc-series/nh-series/NH-76000/NH-76391.html.

Figure 6.2. US Department of the Navy, "NH 76387 Confederate Ironclad *Nashville*," Naval History and Heritage Command Photo Archives, http://www.history.navy.mil/our-collections/photography/numerical-list-of-images/nhhc-series/nh-series/NH-76000/NH-76387.html.

Figure 7.1. Courtesy of Holcombe, 2012.

Figure 7.2. US Department of the Navy, *Official Records of the Union and Confederate Navies in the War of the Rebellion* (1906; repr., Harrisburg, PA: National Historical Society, 1987), ser. 1, 21:663.

Figure 8.1. US Department of the Navy, "NH 76384 CSS *Albemarle* (1864–1864)[*sic*]," Naval History and Heritage Command Photo Archives, http://www.history.navy.mil/our-collections/photography/numerical-list-of-images/nhhc-series/nh-series/NH-76000/NH-76384.html.

Figure 8.2. Still, *Confederate Shipbuilding*, 22.

Figure 8.3. Courtesy of Holcombe, 2012.

Figure 8.4. Ibid., 2013.

Figure 9.1. P. Kean, *Report of Evidence Taken before a Joint Special Committee of Both Houses of the Confederate Congress to Investigate the Affairs of the Navy Department* (Richmond, VA: Geo. P. Evans, 1863), 159.

Figure 9.2. US Department of the Navy, "NH 48026 CSS *Jackson*," Naval History and Heritage Command Photo Archives, http://www.history.navy.mil/our-collections/photography/numerical-list-of-images/nhhc-series/nh-series/NH-48000/NH-48026.html.

Figure 9.3. John L. Porter, "Light Draft Iron Clad Steamer, Two Propellors," plan file CX1, collection no. 243, Savannah Squadron Papers, 1862–1865, Manuscript, Archives, and Rare Book Library, Robert W. Woodruff Library, Emory University, Atlanta, GA.

Figure 9.4. Courtesy of Holcombe, 2012.

Figure 9.5. Ibid.

INDEX